BIOLOGY IN PROFILE

Other related titles of interest

BAUM
The Biochemist's Song Book

DUNCAN & WESTON-SMITH
The Encyclopedia of Ignorance

KORNBERG *et al.*
Reflections on Biochemistry

BIOLOGY IN PROFILE

A GUIDE TO THE MANY BRANCHES OF BIOLOGY

Edited by

P. N. CAMPBELL

Courtauld Institute of Biochemistry
The Middlesex Hospital Medical School
London, England, UK

SPONSORED BY THE COMMITTEE ON THE TEACHING OF SCIENCE
OF THE INTERNATIONAL COUNCIL OF SCIENTIFIC UNIONS

PERGAMON PRESS

OXFORD · NEW YORK · TORONTO · SYDNEY · PARIS · FRANKFURT

U.K.	Pergamon Press Ltd., Headington Hill Hall, Oxford OX3 0BW, England
U.S.A.	Pergamon Press Inc., Maxwell House, Fairview Park, Elmsford, New York 10523, U.S.A.
CANADA	Pergamon Press Canada Ltd., Suite 104, 150 Consumers Rd., Willowdale, Ontario M2J 1P9, Canada
AUSTRALIA	Pergamon Press (Aust.) Pty. Ltd., P.O. Box 544, Potts Point, N.S.W. 2011, Australia
FRANCE	Pergamon Press SARL, 24 rue des Ecoles, 75240 Paris, Cedex 05, France
FEDERAL REPUBLIC OF GERMANY	Pergamon Press GmbH, 6242 Kronberg-Taunus, Hammerweg 6, Federal Republic of Germany

First edition 1981

British Library Cataloguing in Publication Data

Biology in profile.
1. Biology
I. Campbell, P. N. II. International Council of Scientific Unions. *Committee on the Teaching of Science*
574 QH307.2 80-42243

ISBN 0-08-026846-3 (Hardcover)
ISBN 0-08-026845-5 (Flexicover)

In order to make this volume available as economically and as rapidly as possible the author's typescript has been reproduced in its original form. This method unfortunately has its typographical limitations but it is hoped that they in no way distract the reader.

Printed in Great Britain by A. Wheaton & Co. Ltd., Exeter

FOREWORD

Bureaucrats like to divide up science into the three neat compartments of Physics, Chemistry and Biology. Anyone who has tried to organize scientific activities within such neat compartments is struck by the oddities of Biology as compared with Physics and Chemistry. The latter two give the appearance of being well organized scientific disciplines catering for the activities and needs of clearly defined professions. The fact that the reality is not quite like this does not prevent the Physicist and Chemist looking askance at Biology which seems to represent a rabble of people with interests as varied as beekeeping and the mass production of antibiotics. We biologists are very varied not only in our interests but also in our involvement in our science. Some may be interested in biology as a hobby and seek no remuneration, whereas others try to make an honest living from it.

Biologists are not ashamed of their oddities, in fact they find the subject fascinating. Those of us who started our working lives as chemists rejoice in the good fortune that led us to biology. It would be foolish nevertheless if we did not acknowledge that the proliferation of biological sciences has its draw-backs. Unlike the physicists and chemists we cannot speak with one voice. But even if we speak with many tongues we can at least try to explain ourselves. What of the bright school children who may be attracted to biology; how are they to choose for further study from among all the varied branches of biology?

These ideas were at the back of our minds when the members of the Committee on the Teaching of Science met in Paris in March 1979. Our Committee is a sub-committee of the International Council of Scientific Unions, which receives support from UNESCO for the improvement of science teaching. Our Secretary, John Lewis, who is a physicist and a schoolmaster, and our Chairman, Charles Taylor, who is a Professor of Physics, welcomed the idea and so I set to work.

The plan has been to selected about 20 biological subjects, all of which can be studied at Universities, and to ask men who are distinguished in these subjects to write brief profiles. I have asked the authors to think of their audience as typically a group of intelligent seventeen-year-olds who are studying biology but who are not too sure which particular branch will satisfy their curiosity and give best scope to their particular talents. I have suggested that a profile should give some background, describe present activities and indicate the future prospects of the subject. Moreover I have suggested that a somewhat personal approach would be welcome. The hope is that the group of profiles will be of interest not only to the students but also to other scientists who are not practising biologists. I think that there is something of interest and of profit for all readers, whatever their scientific experience, for even the biologists do not always appreciate the differences between the various subjects.

The authors have tried to avoid a parochial approach and in choosing the contributions I have attempted an international spread. In this I have not been wholly successful since for various reasons not all my invitations were accepted. Indeed it could be said that the list of authors in large measure represents my friends. It is true that I have worked with many of them and as ever am most grateful to them for their ready cooperation. The style and length of the profiles vary. But my chief hope is that this book will do something to illustrate the rich variety of experience which can be found within the field of biology.

When I was deciding the order of the profiles it struck me that some are concerned with branches of biology which are largely descriptive, others with the applications of biological science to medicine and human welfare, and others depend for their understanding on a knowledge of chemistry and physics. I have therefore arranged the profiles in an order that reflects these different approaches for I thought that the more descriptive profiles would be easier to read and would present the biological phenomena that the later profiles set out to explain.

A reading list is included at the end of those profiles where it is thought that it may be helpful.

I am most grateful not only to my authors but to my numerous colleagues who have looked over the manuscripts and offered their views. I am particularly grateful to Mr J.A.H. Porch who from his experience of teaching has made many suggestions concerning the suitability of the profiles for use in schools.

P.N. Campbell

CONTENTS

The Contributors ix

Profile No.

1 Zoology 3
2 Botany 9
3 Microbiology 15
4 Physiology 21
5 Ecology 31
6 Ethology 37
7 Psychology 43
8 Parasitology 49
9 Pharmacy 55
10 Pharmacology 61
11 Toxicology 67
12 Nutrition 73
13 Food Science 79
14 Endocrinology 87
15 Immunology 95
16 Genetics 101
17 Virology 109
18 Biophysics 115
19 Biochemistry 120
20 Molecular Biology 126

THE CONTRIBUTORS

Zoology	**Professor R. McNeill Alexander,** Department of Pure and Applied Zoology. Barnes Wing, University of Leeds, Leeds LS2 9JT, U.K.
Botany	**Professor Harold W. Woolhouse,** Director, John Innes Institute, Colney Lane, Norwich NR4 7UH, U.K.
Microbiology	**Dr Brian Bainbridge,** Department of Microbiology, Queen Elizabeth College, Atkins Building, Campden Hill, London W8 7AH, U.K.
Physiology	**Dr M.C. Shelesnyak,** The American Physiological Society, 9650 Rockville Pike, Bethesda, MS 20014, U.S.A.
Ecology	**Dr Malcolm J. Coe,** The Animal Ecology Research Group, Department of Zoology, South Parks Road, Oxford OX1 3PS, U.K.
Ethology	**Dr Peter J.B. Slater,** School of Biological Sciences, The University of Sussex, Falmer, Brighton, Sussex, BN1 9QG, U.K.
Psychology	**Professor H. Gwynne Jones,** Department of Psychology, University of Leeds, Leeds LS2 9LS, U.K.

Parasitology

Professor Francis E.G. Cox,
Department of Zoology,
King's College,
Strand,
London WC2R 2LS, U.K.

Pharmacy

Professor Sir Frank Hartley,
(Formerly School of Pharmacy,
University of London);
now 146 Dorset Road,
Merton Park,
London SW19 3EF, U.K.

Pharmacology

Professor George Brownlee,
Department of Pharmacology,
King's College,
Strand,
London WC2R 2LS, U.K.

Toxicology

Dr W. Norman Aldridge,
MRC Toxicology Unit,
MRC laboratories,
Woodmansterne Road
Carshalton,
Surrey, SM5 4EF, U.K.

Nutrition

Professor Arnold E. Bender,
Department of Food Science & Nutrition,
Queen Elizabeth College,
Campden Hill Road
London W8, U.K.

Food Science

Professor Alan G. Ward.
(formerly of the Department of Food and Leather Science, University of Leeds);
now 35 Templar Gardens,
Wetherby, W. Yorks, LS22 4TG, U.K.

Endocrinology

Dr Bryan Hudson,
H. Florey Institute,
University of Melbourne,
Parkville,
Victoria 3052, Australia.

Immunology

Professor Peter N. Campbell,
Courtauld Institute of Biochemistry,
The Middlesex Hospital Medical School,
Mortimer Street,
London W1P 7PN, U.K.

Genetics

Professor Joshua Lederberg,
President, Rockefeller University,
1230 York Avenue,
New York, NY 10021, U.S.A.

Virology

Dr Erling Norrby,
Department of Virology,
Karolinska Institutet SB1,
S-10 S/21 Stockholm, Sweden.

Biophysics

Professor David M. Blow,
The Blackett Laboratory,
Imperial College,
London SW7 2BZ, U.K.

Biochemistry

Professor Emil L. Smith,
Department of Biological Chemistry,
UCLA School of Medicine,
The Center for the Health Sciences,
Los Angeles, CA 90024, U.S.A.

Molecular Biology

Professor Peter N. Campbell,
Courtauld Institute of Biochemistry,
The Middlesex Hospital Medical School,
Mortimer Street,
London W1P 7PN, U.K.

ZOOLOGY

Profile No. 1

ZOOLOGY

by

Professor R. McNeill Alexander
(Professor of Zoology in the University of Leeds)

Zoology includes all aspects of the scientific study of animals. Consequently, it overlaps many of the other disciplines described in this book. Many branches of zoology can equally well be described as ecology, endocrinology, ethology, evolutionary biology, parasitology, pharmacology or physiology. The more interested a scientist is in the lives of individual animals, the more likely he is to call himself a zoologist. For instance, many scientists have performed experiments on the giant nerve fibres possessed by squids. Some of them have been investigating the mechanism by which nerves transmit messages, and have chosen squid nerve fibres because they are the largest available. These scientists would probably describe themselves as physiologists. Other scientists have studied the same nerve fibres with the aim of understanding why they are so big, and their role in the lives of squids. These scientists would call themselves zoologists.

Some zoologists enter the subject through natural history. They have been keen bird watchers, or insect collectors, and see zoology as an extension of their hobby. They may be disappointed, because most zoology is very different from most natural history. A man who watches birds at their nests and tries to see what food they bring to their young would probably call himself a naturalist. Another who attaches oxygen analysis equipment to nest boxes so that he can estimate the energy consumption of the young birds would probably call himself a zoologist.

The first thing you need if you want to become a zoologist is an interest in animals. The second is an ability to cope with a wide range of scientific disciplines. An all-round zoologist would have to be competent in many different branches of chemistry, physics, mathematics and engineering. He would have to be familiar with the concepts and techniques involved in the study of problems ranging from the molecular structure of an insect hormone to the decline of the world population of whales. Unfortunately, there are very few all-round zoologists.

I will try to convey some of the variety and interest of modern zoology by describing a few of the investigations and discoveries of the 1970s.

Lions live in prides which typically consist of two or three adult males, five to ten adult females and some cubs. The members of the pride cooperate in hunting and there is remarkably little rivalry between the males. How has this social system evolved? To answer this question it was necessary to learn to recognize individual lions, and to keep a close watch on two prides in Tanzania for seven years. It was discovered that all the females in a pride have been born into the pride, and so tend to be closely related and have a lot of their genes in common. The males all come from another pride and probably share many genes. Evolution tends to favour cooperation in family groups because your sister's or even your cousin's children share many of your genes, and by helping them you increase the probability that genes identical to your own will be transmitted to future generations. Every few years a group of young males, born in another pride, drive out the old males and take their place, often killing their cubs. There is no evolutionary advantage for them in supporting cubs which are unrelated to them.

Paramecium is a protozoan (a single-celled animal), about the size of the full stop at the end of this sentence. Since it is only a single cell, it has no nervous system. It swims actively, and if it collides with an obstacle it reverses, turns slightly and advances again. If this fails to get it past the obstacle it tries again. How can an animal without a nervous system behave in such a complex way? Ingenious and delicate experiments have provided an answer. In some of the experiments fine electrodes were pushed into *Paramecium* and used to record the changes of electrical potential which occur when the animal is poked (and presumably when it collides with something). In others, the outer membrane of the animal was made permeable to salts by treatment with a detergent, and it was shown that the animal swam forwards in some solutions and backwards in others. It was concluded that when *Paramecium* collides with an obstacle, changes occur in its outer membrane which allow calcium ions to diffuse in and stimulate reversal of swimming. These changes resemble the changes (involving sodium instead of calcium) which occur in the nerve cells of other animals when they are stimulated. The whole body of *Paramecium* behaves like a nervous system.

It has been suggested that dinosaurs may have been warm-blooded like mammals and birds, not cold-blooded like other reptiles. Much of the evidence seems to me unconvincing, but one line of argument deserves serious consideration. Because of the faster metabolism associated with their warm-bloodedness, mammals and birds need five or more times as much food as reptiles of similar size. Consequently, a population of warm-blooded carnivores needs a much larger prey population to support it, than does a similar population of cold-blooded carnivores. An exceptionally rich deposit of fossils in Canada yielded 22 carnivorous dinosaurs, 246 (mostly larger) herbivorous dinosaurs on which they presumably preyed, and no other large animals. If these numbers reflect the proportions of predators and prey in the living population, a relatively small population of predators was feeding on a much larger population of prey. Ecological calculations based on these proportions admit the possibility that the dinosaurs were warm-blooded.

The pink bollworm caterpillar is the most important pest of cotton crops in many countries. The adult female moths attract males by releasing a volatile organic compound with a strong odour. This suggested a novel method of pest control. Small quantities of a synthetic compound, almost identical with the natural sex attractant, were distributed over the fields. The odour spread everywhere and the males had difficulty in finding the females. In the initial trials the number of bollworm caterpillars in treated fields at the end of the season was less than 10% of the number in a neighbouring untreated field.

Encarsia is a tiny insect with wings only 0.5 mm long. High-speed cine films of it flying showed that it made 240 complete cycles of wing movements per second, but conventional aerodynamic calculations led to the false conclusion that this was not fast enough to enable *Encarsia* to fly. The films showed that the wings are clapped back-to-back at the top of each upstroke and then separate again, starting with their front edges. New calculations showed that this would set up a circulation of air around the wings, sufficient to support the insect's weight. *Encarsia* uses an aerodynamic principle which was previously unknown to engineers.

Human divers who make deep or long dives are liable to a serious condition called the bends, unless they adopt a slow decompression routine.

The high pressure at depth makes nitrogen from the lungs dissolve in the blood. When the diver surfaces, nitrogen comes out of solution and forms bubbles in blood vessels and nervous tissue, causing pain and even death. How do whales avoid the bends? Two dolphins were trained to dive on command to a submerged bell-push, ring the bell and return to the surface. After a series of dives to 100 m they were lifted into a boat and a hypodermic needle connected to a mass spectrometer was pushed into the muscles of the back. The instrument showed that the nitrogen content of the muscles was 2-3 times the normal value. The rate of loss of the extra nitrogen was monitored over the next 30 minutes. It was calculated from this rate that if the blood had had free access to the air in the lungs during the dives, the nitrogen content of the blood and muscles would have risen to 4-5 times normal. It was suggested that the lungs collapse, driving the air into the windpipe where it has little opportunity for diffusion into the blood, during the deeper part of the dive. This helps to prevent the nitrogen content of the tissues from rising to dangerous levels. There is other evidence to support the hypothesis.

The Waddenzee is an area of tidal inlets and mud flats between the Dutch coast and a chain of islands. It was proposed to build dams between the islands, cutting it off from the North Sea, and then to drain it. It was suspected, however, that the Waddenzee might be an important nursery area for young fish and that drainage might have a serious effect on North Sea fisheries. Four fishing vessels were employed to survey the populations of young fish in and around the Waddenzee, using trawls fitted with shrimp nets. It was estimated from the catches that the Waddenzee contained 60-80% of the young plaice and soles in Dutch coastal waters. The estimated numbers of young fish seemed to be less than the area was capable of supporting but it was feared that drainage of even part of the Waddenzee might be very damaging to the plaice and sole fishery.

These few examples will give some impression of the variety of topics which are studied by zoologists. An even wider range of topics would have had to be illustrated, were it not that several branches of zoology (animal ecology, ethology and parasitology) are treated in this book as separate subjects.

A zoologist should respect the lives of the animals he studies. Is an experiment justifiable, if it involves killing animals? The answer depends on the quality of the experiment and the rarity of the animals. If the animals are uncommon it may well seem wrong to proceed, even if the experiment seems capable of making an exciting contribution to knowledge. If the animal is rare it may be protected by law and the zoologist will not normally be permitted to make the experiment.

Moral questions also raise concerning experiments on living animals. Is it justifiable to carry out possibly painful procedures or surgical operations on living animals? The answer depends on the severity of the procedures and on the value (academic or practical) of the knowledge it is hoped to gain. In Britain, experiments on living vertebrate animals are very strictly controlled by law. Every experimenter needs a licence from the Home Office. By itself, this permits only experiments in which the animal is under anaesthetic throughout and is not allowed to recover consciousness. Additional authority is needed if it is to be allowed to recover or if an even slightly painful procedure without anaesthetic is proposed. Additional authority is also needed for experiments with monkeys (because they are quite closely related to man) or with cats, dogs or horses (because people feel particularly strongly about species which they keep as pets).

Some people imagine that zoologists work in zoos. A few of them do, but far more are employed in other professions, for instance in the pesticide and pharmaceutical industries, in research and advisory work in agriculture and fisheries (including fish farming), in wildlife conservation and pollution research, in museums, schools and universities and in television companies producing wildlife programmes.

BOTANY

Profile No. 2

BOTANY

by

Professor Harold W. Woolhouse
(Professor of Botany in the University of Leeds. Now appointed Director of the John Innes Institute, Norwich, and Professor of Biology in the University of East Anglia)

The scope of botanical studies

Botany is simply defined as the study of plants. In practice it is a vast subject which embraces many topics. These include evolution (as studied from both the fossil record and from cytogenetics and breeding work with living plants); taxonomy and the geographical distribution of species both at the present day and in times past. Botany also covers the study of plant growth, development and reproduction and the physiological and biochemical mechanisms by which these processes are sustained; ecology – the study of plant communities and their environmental relationships and lastly, though arguably most important of all, the application of many of these lines of work to the improvement of yields and quality in crop plants. There is another vantage point from which we can readily appreciate the scope of botanical studies, and this is by considering the diversity of plant life. We may consider, for example, the range of size and life forms in plants, from giant trees of the rain forest down to the micro algae, invisible to the naked eye, which compose the phytoplankton – the primary producers of the sea. Another way of emphasising the diversity of plants is to think of the many different groups, the fungi, algae, mosses, ferns, conifers and flowering plants – all of which are the botanist's concern.

Employment prospects in botany

I have deliberately dwelt at length on the vastness of my subject in order to show how this gives rise to a great range of opportunities for the employment of botanists. Thus it is that students trained in selected aspects of the biology of plants may subsequently find employment in such widely ranging enterprises as crop science and forestry, marine biology, ecology, conservation, industrial and university-based research and in teaching, to name but a few of the more obvious.

It is a curious fact that, despite the wide scope of botanical work, the consequent range of opportunities of employment and the undoubted intrinsic interest of many aspects of the subject, the recruitment of students into the profession is low relative to the number of openings. Much soul searching has gone into trying to explain this state of affairs. Some have suggested that prospective students are discouraged by the emphasis on classification which started with early herbalists and persisted into the botanical exploration of the nineteenth century; but it is not true of most modern university courses. Others have sought the explanation in the approach to biology in schools, where teachers find it easier to build on the greater initial excitement aroused by animals because of their capacity for intelligent behaviour and rapidity of response to stimuli. Yet again it often happens that some students become discouraged even at a more advanced level of research. For example in biochemical work, the existence of a rigid cell wall and the presence of many phenolic and other toxic compounds tends to impede progress and has led to disillusionment with plants. This may in part explain why, even today when most of these difficulties have been overcome, one may search in vain in most of our university biochemistry departments for anyone whose interests centre on the study of plants.

It may appear at first sight that I paint a gloomy picture; but to the discerning newcomer, faced with a choice of subject, the brighter side to this situation should soon become apparent. Thus as we consider many of the aspects of biology, particularly veterinary science and medicine, we find that supply greatly exceeds demand and entry into the profession is difficult. In plant science and some of the related aspects of biology, on the other hand, the opportunities are there: they also have a sociological appeal, for basic training can lead to the study of the exciting problems waiting to be solved in many aspects of human affairs, varying from food supplies to conservation, and from timber production to the farming of the seas. Let us then survey some of the growth areas of botanical science and seek to identify challenging problems which they offer.

Growth areas in botanical research

The distinctive feature of most plants is their capacity to carry out photosynthesis. In their photosynthetic apparatus plants possess the only significant system on earth for converting the energy of sunlight into the energy of chemical bonds. This attribute gives plants a unique importance as the energy source on which ultimately depends the functioning of all the living organisms of oceans, forests, prairies, deserts and agricultural lands. Although enough is known about photosynthesis to fill books and courses of lectures, a multitude of problems remain. For the scientist with a flair for fundamental problems there is surely a Nobel Prize to be won for the elucidation of the way in which plants are able to split water molecules and so obtain the electrons needed to assimilate carbon dioxide into carbohydrate. But it now becomes apparent that, even when such basic issues in the photochemistry and biochemistry of photosynthesis are understood, there are still hosts of problems which remain. These centre on how plants are able to modify their photosynthetic apparatus to function under particular environmental conditions. For example, we now know that the optimum temperature for photosynthesis in some species is as low as 10–15°C, whilst in others it is as high as 40–45°C; moreover, species in the latter category may cease photosynthesis altogether at 10–15°C. Other species possess adaptations enabling them to sustain photosynthesis under conditions of nutrient deficiency, water stress, salinity or in deep shade, which would prove lethal for most species. Probing the nature of these adaptations leads us into all manner of problems concerned with the structure and stability of membranes, altered properties of enzymes and variations in cellular organisation.

For those whose preferences run more in the direction of developmental biology, plants provide some particularly intriguing questions. A brief consideration of two topics, the role of light and the part played by hormones in plant development, will illustrate the problems. It has long been known that the intensity and spectral composition of light may profoundly alter many of the morphogenetic responses of plants, as for example the elongation of stems, the unfolding of leaves, the induction of flowering and the nocturnal movements of leaves and petals. Exciting progress in this field has led to the isolation of a photoreversible pigmented protein which functions as the receptor for the photostimulus; we know just a little of where some of this pigment is located in the cell but stand in almost total ignorance of how it functions to elicit the morphogenetic changes.

Research on the mechanisms of hormone action in plants offers what promises to be one of the really tough problems in modern biology. It is now widely known that animals produce a large number of hormones which vary greatly in chemical structure and in their modes of action, yet in higher plants only five hormones are known, although three of these exist in a host of minor variant forms. The curious fact is that many stages of development in plants require the concerted action of these hormones in various combinations; moreover the action of a given hormone on a cell changes in the course of the life of the cell. We have yet to determine the nature of the hormone receptors in plant cells or how they act to elicit specific responses; harder still, but of crucial importance, will be to determine the subtle changes in the cell which enable it to change the way in which it responds to hormonal stimulation at successive stages of its development.

Work on cancer, immunology and the compatibility of grafted tissues has led to a great upsurge of interest in processes by which animal cells are able to recognise and respond to one another by acts of mutual identification as self or non-self. No less an explosion of interest in these problems is to be found amongst botanists; how, for example, does it arise that amongst the multiplicity of wind-pollinated species a stigma will accept only the pollen of its own kind; why can some species be used as rootstocks for the grafting of fruit trees while other species will not accept the grafts; how does it arise that the gamut of brown seaweeds shedding their naked eggs and free-swimming sperms into the sea, mate like with like and rarely hybridize; or yet again why many fungal pathogens are found only in particular host species or even in a particular genotype of the host? In all these cases, as in animal cells, macromolecules located on the surfaces of the cells determine the recognition of self, own species, or alien visitor. How these molecules function to instruct the cell to an act of acceptance or rejection raises fascinating possibilities for further experiment.

While we are considering particularly recent developments, it is worth noting how rapidly the techniques of cell fusion, tissue culture and genetic engineering are finding practical application in the plant sciences. The most abundant protein on earth is the enzyme which fixes carbon dioxide in green leaves. This enzyme has an amino acid composition well-suited to the nutrition of higher animals. The genes responsible for the CO_2-fixing enzyme have now been isolated and hopes are high that they may be cloned and perhaps induced to direct the synthesis of the protein in the cells of micro-organisms. Similar progress is being made with the storage proteins of cereals and legumes which are vital to human nutrition. For the intending molecular engineer amongst plants, there exists the exciting challenge of increasing the dosage of the storage protein genes and, at best, the quantity of their products in the crop species from which they were derived.

Many people's earliest interest in biology is aroused by the observation of plants and animals in the open air. There is a sense in which work in the field remains the highest form of biology, or perhaps we might say its culmination, for it is in the field that the analytical studies must ultimately be brought together to account for the functioning of the intact organism as part of a community or ecosystem. Ecologists begin by seeking to understand the factors which govern the composition, relative stability and turnover of the components, that is to say the structure and dynamics of an ecosystem. They then seek to incorporate knowledge gained from the analytical fields to explain the observed dynamics. Increasingly the ecologist has to face severe examination of his findings as he is required to incorporate them into predictive models to serve the needs of society. Government planners, leisure services and other bodies now require the ecologists' models for such purposes as the prediction of management policies for the conservation of ecosystems of outstanding scientific interest, the maintenance of soil fertility and prevention of erosion, the maintenance of genetic diversity, the retention of desirable visual landscapes and the restoration of areas which have become degraded.

I have endeavoured in this brief note to highlight the breadth of studies encompassed by botanists, and to illustrate something of the challenges and excitements which await the newcomer to botany. I end with a caution. The study of botany is no different from other forms of science in its demands upon time, patience and determination in the face of unsuccessful experiments. Against this there is the advantage of opportunities for many types of student, the collector and classifier, the anatomist and the experimentalist; and for each of these there is the satisfaction when things do succeed, which goes deeper than words, which the present author can best express from personal experience by saying how grateful he is that he became a botanist.

MICROBIOLOGY

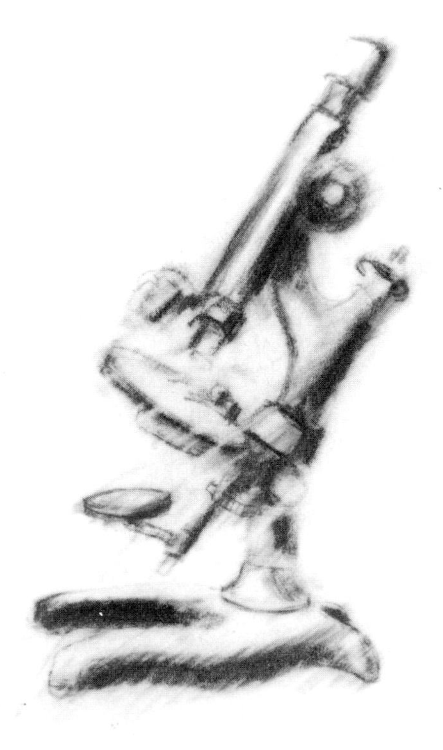

Profile No. 3

MICROBIOLOGY

by

Dr Brian Bainbridge
(Senior Lecturer at Queen Elizabeth College, London)

I am often asked by relatives and friends about the aspects of microbiology that are relevant to everyday life. It is not difficult to list a few of the more dramatic of these, such as antibiotics, infectious diseases, food poisoning, food preservation, water purification and the destruction of man-made materials. However, the scope of microbiology is now much wider than this and man's future is likely to depend to a significant extent on his ability to control and exploit the activities of microbes.

The simplest definition of microbiology is that it is a branch of biology concerned with small forms of life, the microbes. Many of these exist as single cells or small groups of cells which can be seen only under the light microscope or, if viruses, under the electron microscope. This definition, based on size, results in a heterogeneous group which spans the plant and animal kingdoms since it includes the viruses, bacteria, yeasts, micro-fungi, slime moulds, micro-algae and the protozoa. The single feature which links a variety of small forms of life is the common need for specialised techniques to study their activities. Microbes occur almost everywhere from the snows of the Arctic to the hot springs of New Zealand and from the fuel tanks of jet aircraft to the water in your central heating system. Natural habitats often contain such an abundance and variety of microbes that it is impossible to study accurately the activities of a particular microbe without first separating it from the surrounding microbes. To do this it is necessary to have a nutrient jelly, usually agar-based, which is free from all microbes. To achieve this, equipment and all material needed to promote growth are usually heated with steam under pressure in an autoclave or pressure cooker, a process known as sterilisation.

Microbes occur in the air, on the skin and on experimental surfaces so that special precautions, known as sterile techniques, are needed to prevent contaminating microbes from reaching the growth medium. The material from the natural habitat is diluted in such a way that single cells are separately inoculated onto the nutrient agar plate. Each single cell grows to produce a clone or colony of a particular microbe which is now known as a pure culture. Large numbers of almost identical cells can be produced with ease and these are very useful for biochemical and genetic experiments as well as for industrial processes. Any cell which divides to give separate daughter cells can be analysed by these techniques. Microbiological techniques have even been applied to human skin cells, plant cells and insect cells in liquid culture.

The origins of microbiology can be traced back, not surprisingly, to the invention of the light microscope as microbes could not be seen until this became available. Experimental microbiology was established in the 19th century by the pioneer work of Pasteur in France and Koch in Germany. Pasteur showed that microbes were responsible for a number of industrial processes such as the production of wine and beer and he also made a significant contribution to the prevention of spoilage by the introduction of limited heat treatment called, after him, "pasteurization". His work formed the scientific basis of the fermentation industry, which has grown considerably in this century and is an important part of what is now known as "biotechnology". Koch demonstrated conclusively that microbes, particularly the bacteria and later the viruses, could cause disease and this led to the rapid recognition of the bateria responsible for anthrax, tuberculosis and cholera. Medical microbiology developed rapidly and the scientific basis for vaccination, developed empirically by Jenner, was discovered. The next major advance was the disovery of chemicals which could selectively kill microbes. A number of bacterial diseases were cured by the use of these drugs and this has led to the current practice of too optimistically regarding antibiotics as the miracle cure for all diseases. In parallel with this work Winogradsky and others had shown the importance of microbes in the nitrogen, carbon and sulphur cycles. These processes are essential for the recycling of organic and inorganic material in the environment as well as for the maintenance of soil fertility.

In the 20th century a study of microbes has had a major influence on genetics, molecular biology and biochemistry. Our knowledge of the biological functioning of the genetic material, DNA, stems largely from the study of microbes and microbial genetics. This approach has culminated in the spectacular achievements of recombinant DNA technology (genetic engineering). It is now possible to insert a piece of DNA, which represents the gene for a particular protein, into a micro-organism in such a way that it will be replicated as if it were the natural genetic material of the host cell. In this way it has become possible to cause a bacterium to synthesize an animal protein. Techniques are available to permit selection of the particular colonies of bacteria that have been so "transformed". Hence it is possible in principle to "clone" a baterial colony to give a population that contains only the one foreign gene which has been chosen. Moreover the gene may be subsequently recovered from the bacterial clone.

The major applications of microbiological knowledge have been in ecology, agriculture, the food industry, the exploitation of microbial activity, medical microbiology and the pharmaceutical industry. Valuable information can be obtained from pure cultures but many natural environments and some industrial processes are a complex of microbial interactions. Soil, which of course in the main consists of inanimate chemicals, also contains a large, diverse population of microbes which interact in a variety of ways. In particular the microbes interconvert some of the essential nutrients in the soil. Soil can be simulated in the laboratory by the use of columns containing glass beads or clay particles. Known organisms are inoculated onto a column and a solution containing, for example, ammonium ions which are commonly present in soil, is trickled down the column. The effluent can then be analysed for the presence of ammonium, nitrite or nitrate ions and the effects of polluting a field with synthetic chemicals can be simulated by adding the chemical to the solution entering the column. Water purification systems based on activated sludge exploit the interactions between bacteria, fungi and protozoa. Experimental microbiology can simulate one of these interactions by growing a mixture of a single bacterial type and an amoeba. The amoebae feed on the bacteria and complex growth kinetics result. Models for this prey/predator system have been drawn up by computer and used to predict the results actually obtained. Nitrogen fixation in the soil occurs either through the activity of free-living bacteria or by symbiotic bacteria in association with plants, such as legumes. Research in genetic engineering has enabled nitrogen-fixing genes to be transferred between different bacteria and this has raised the possibility of extending the range of symbiotic bacteria so that they can fix nitrogen in association with wheat. Microbes are also important in the production of silage for cattle by the fermentation of plant remains in the absence of oxygen. This produces acids thus increasing the palatability of the plants and also preserving their nutritional value. The digestion of grass by cows and sheep is dependent on the presence of cellulose-digesting bacteria in the rumen. A recent report has described the cow as a "living, mobile, self-reproducing and edible fermenter" to which should be added the adjective "inefficient" as it is subject to disease and is 400 times less efficient than an equivalent weight of microbes growing in a fermenter. An active area of research is in examining the possibility of converting cellulose directly into useful products such as sugars. Plants are the ultimate source of food for man so it is obvious that he should be able to control pests which attack crop plants. Bacteria have been discovered which produce a chemical which is selectively active against insect larvae and this has been very successful in large-scale field trials.

Traditionally microbes have been used for centuries to produce palatable foods, to increase the keeping quality of food and to yield useful food chemicals. Baking and brewing are well known but the production of lactic acid in yoghurt and acetic acid in vinegar are also important. The production of yeast extract has turned a waste product and potential pollutant from the brewing industry into a valuable source of vitamins. More recently microbes have been grown to produce a direct source of food for animals or man. Bacteria and yeast have been grown on methane, methanol, paraffins or on wastes from agriculture or industry. Bacteria grown on methanol obtained by oxidation of the methane in natural gas from the North Sea are now being marketed as an animal fodder, and yeast grown on ethanol is being used as a food additive for humans. These processes are dependent on cheap sources of organic substrates from the oil industry or on waste products from the processing of plant materials. It should be realised, however, that it is very likely that oil was produced by the action of microbes on decaying plant materials in the first place. Micro-algae have recently been investigated as a potential source of food and chemicals and this raises the possibility of using the energy of the sun more efficiently than in current agricultural practices. Food from any source can be a source of disease either from food poisoning, due to toxins derived from bateria or fungi, or from food-borne infection due to the presence of pathogenic bacteria. A long term aim of food microbiology is to devise methods of processing to produce stable foods that will not deteriorate.

The metabolic activities of microbes have been exploited in a variety of ways. The fermentation industry has produced, for example, ethanol, citric acid, acetic acid, lactic acid and glutamic acid, the last an important food additive. The major factor here is whether the microbiological or the chemical process is the cheaper. Microbes exist which can attack virtually every naturally ocurring organic compound and this has been used in systems for purifying domestic, industrial and agricultural effluents. Phenols and other toxic chemicals can be degraded but newly synthesized chemicals are often not susceptible to attack, a situation which was highlighted by the persistence of DDT and some detergents. New detergents have since been found which are "biodegradable". A useful by-product of sewage disposal is the generation of methane which can be burned as a fuel to generate electricity. Other bacteria can generate hydrogen and this may be the fuel of the future as it is pollution-free, although there are problems because it is explosive in mixtures with oxygen. Genetic techniques have been used to produce a bacterium which can degrade oil and oil derivatives and it has been suggested that this may be a feasible way of dispersing oil pollution at sea without the damaging and expensive use of detergent sprays. Many countries are trying to find alternative sources of energy to reduce dependence on oil and Brazil is using yeast to produce ethanol from sugar cane for use as a fuel in cars. A further use of microbes has been the application of bacteria to recover metals such as iron and uranium from insoluble ores by leaching and also to extract heavy metals from dilute solutions as part of water purification. Bacterial exopolysaccharides and gums have been used on a massive scale, some 36,000 tonnes per annum, to aid in the recovery of oil from porous rocks. The chemical is pumped into the rock strata below the oil bearing rocks where it makes the water more viscous, providing a firm pillow which aids in the recovery of the oil. Similar compounds are used in savoury sauces and non-drip paints.

Much of medical microbiology is now well established but there have been some exciting new trends. Antibiotic production by filamentous fungi and by the filamentous bacteria, the streptomycetes, is well known but we now have the ability to improve strains using recombinant DNA technology. The use of antibiotics in medicine and animal husbandry has resulted in high levels of bacteria resistant to these antibiotics. Plasmids carrying resistant genes can spread from harmless bacteria to dangerous pathogenic bacteria. Sensible use of antibiotics is therefore essential. An important development from this work has been the exploitation of plasmids for genetic engineering. Genes from virtually any organism can be grafted onto the plasmids and returned to a bacterial cell. This should enable effective harmless vaccines to be produced against viruses such as hepatitis B and influenza. Considerable progress has also been made towards the synthesis of human insulin and other human hormones in bacterial cells. This raises the possibility of the industrial production of cheap hormones using microbial fermentation methods. There has been considerable controversy over the possible hazards of this work but at present the risks seem to have been exaggerated.

Microbes have also been used as a source of enzymes for medical use, e.g. asparaginase for the treatment of leukaemia, for domestic use, e.g. proteases in washing powder, and for industrial conversions, e.g. glucose isomerase to change glucose into the sweeter fructose. Another interesting development is to attach enzymes to columns of inert material. In this way the enzyme is immobilized. A solution containing a substance which is a substrate for the enzyme is passed down the column and the new product collected in the effluent. The product of the enzyme reaction is thus obtained free from contamination by the enzyme and the valuable enzyme can be re-used. An immobilised enzyme system has been used industrially to deacylate benzyl penicillin in the production of semi-synthetic penicillin.

Future developments in microbiology are likely to be very diverse but it seems probable that the major pressures will come from man's requirement for energy, food and protection from diseases. Microbes have an impressive range of biochemical abilities which have been used in biotechnology. It seems very likely that a significant number of man's problems will be solved by the exploitation of microbes.

Further reading

Dixon, B. (1976) *Invisible Allies: Microbes and Man's Future*. Temple Smith, London.

Camera Talks Ltd. Tape/slide presentations on the impact of microbes on man. Titles include: *Mushrooms*, *Penicillin*, *Viruses*, *Sewage*, *Food from Microbes and Methane*. Camera Talks Ltd., 31 North Row, London W12 3EN.

Noble, W.C. and Naidoo, J. (1979) *Microorganisms and Man*. Studies in Biology, No. 111. Edward Arnold, London.

Postgate, J.R. (1975) *Microbes and Man*, 2nd ed. Penguin Books, Harmondsworth, Middlesex.

Reid, R. (1974) *Microbes and Men*. BBC Publications, London.

PHYSIOLOGY

Profile No. 4

PHYSIOLOGY

by

Professor M.C. Shelesnyak

(Executive Editor of "The Physiology Teacher" section of "The Physiologist")

Imagine sitting, deeply engrossed in reading, in a quiet room, dark except for the lamp over your shoulder. Imagine you are reading a thriller, a suspense espionage story. It is past midnight; all is very quiet. Suddenly there is the squeal of a dry hinge as the door opens; or the explosion of an engine back-firing. You know what happens: your startled body jerks upright, your heart "pounds" as the pulse increases; your breathing pattern changes. You begin to sweat. The alarm reaction may even relax your bladder or rectal control. These are bodily reactions which all of us have experienced.

Stop and think for a moment what started this whole chain of events: a miniscule amount of physical energy – the few decibels of sound from the squeal or explosion noise. The displacement of particles in the air (sound) caused the eardrum to vibrate, transmitting the motion via three small bones of the middle ear to the inner ear, where the cochlear nerve transfers the information to the brain by electrochemical means. The brain "reaction" results in sending messages throughout the body; most generally known is the release of adrenalin from the adrenal gland (at the top of the kidney) and the network of autonomic nervous system ganglia. There are also changes in blood fat and blood sugar levels, changes in body heat production. All of these responses involve functional changes in or about body cells – physiological, biochemical, structural.

The discovery, study and understanding of the almost infinite changes, large and small, which take place in the living organism in response to changes in the external and internal environments, are the physiologists' world.

Physiology is the study of the processes of living nature. Its scope includes everything we classify as living as opposed to nonliving. As a scientific discipline, physiology is, in certain respects, both the oldest and the youngest of scientific endeavours. It aroused the curiosity of ancient man. By the 4th century B.C., this curiosity had evolved into systematic scientific investigation. Late in that century, Aristotle painstakingly studied animals and recognized and classified the relation of structure and function in many species. Physiology is, on the other hand, perhaps the youngest of all sciences in terms of vigour, unexplored frontiers, and exciting prospects.

The aim of physiology is to probe biological processes and to gain a detailed understanding of the underlying physical and chemical mechanisms. Physiology is concerned not only with *what happens* but also *how and why it happens*. Its vistas include the properties of subcellular structures and how these affect cell function; the relation between cell function and the function of a tissue or organ; the interaction among organs necessary for survival of the organism; and the relationship of the organism to its variable and often hostile external environment.

It is inevitable in science that virtually every discovery discloses another mystery, another puzzle piece without which the picture is incomplete. And even while we are expanding our knowledge of man on his native planet, his world is expanding and with it the boundaries of physiology. As man moves into new environments, new problems arise and demand resolution. How will man fare living for long periods of time in weightless space when gravity has influenced his every function and action from the time of birth? As men venture deeper under the sea and stay there longer, new questions about physiological hazards and safety arise. At what depth will pressure effects cause unacceptable changes in tissue structure? Are there residual effects of repeated exposure to great depths?

All creatures have a highly pronounced daily rhythm, and man is no exception. This rhythm is often synchronized with external cues, particularly light and dark. Physiological changes follow a relatively precise pattern throughout the day. The human body temperature, blood pressure, respiration, and hormone levels all rise and fall, ebb and flow predictably. The light-dark cycle in Earth orbit is only 90 minutes long. Jet aircraft passengers and crews flying through time zones have a similar, although less extreme, experience. While the passenger may have an opportunity to sleep off the effects, this luxury is not afforded to the flight crew. What happens to man's performance when familiar external cues are absent or radically altered?

Each year, for many years, we have dumped millions of tons of man-made waste products into the air. Until corrective effects are successful, this pollution will sicken man and kill flowers and crops. How will this affect the ecological balance? And what influence will it have on the survival of organisms up and down the phylogenetic scale? These are just a few of the problems of modern life that have added to the scope of physiology.

The breadth, depth, and complexity of the problems to be resolved in physiology offer exciting and important challenges to individuals with a broad spectrum of talent and a wide range of interest and educational background. The success of physiology to date has been largely the result of its ability to attract these individuals and to focus their interests on life processes. For example, physiologists with a talent for physics, mathematics, zoology, physical chemistry, and electrical engineering have all contributed to our understanding of how a painful stimulus is transmitted to and processed by the brain.

Many of the remaining problems in physiology are vast – far more complex even than the discovery of the mechanism of nerve impulse – and exceedingly demanding. The physiologist of the future will not be defined so much by his educational background and talents as by the direction toward which he focuses these. If you are gifted in mathematics, engineering, physics, chemistry, or biology, physiology can accommodate you and direct these skills toward a productive exploration of life processes.

The problems will not be easy to solve because experiments on living systems can never be as precisely controlled or executed as those on man-made systems. But successes *are* achieved and more is learned each day about disease, man and his environment, behaviour, and normal function.

Physiology – past, present and future

Curiosity about the operation of living things was the origin of physiology. The name comes from the Greek words *physis*, meaning nature, and *logia*, meaning knowledge. Aristotle and the *physiologoi* or physiologists of ancient Greece who first related structure to function in plants and animals were natural scientists in the broadest sense. Galen, a second century physician, is considered one of the founders of experimental physiology. Most of his notions proved to be quite wrong, but his teachings were followed slavishly for more than a thousand years. Finally, the Renaissance broke through the stifling authoritarianism of the Middle Ages and permitted new knowledge to grow.

At first, progress was slow. Many advances in physiological knowledge had to await discoveries in physics and chemistry. In 1628, Sir William Harvey published the first breakthrough in physiological knowledge in 1300 years. He demonstrated that the heart pumped blood through the circulatory system of living animals. During the 19th century, physiological knowledge took another quantum leap. The impetus came from the work of a Frenchman, Claude Bernard. Bernard demonstrated that the body breaks down complex chemicals in order to use them and that it builds and stores chemicals. This work contributed to our understanding of metabolism. His greatest contribution was a description of an essential characteristic of life – the concept that organisms preserve internal stability despite external change.

With advances in chemistry, physics, mathematics, and technology, there has been an explosion of physiological knowledge in the 20th century. As an example I may mention the great Russian scientist, Ivan Petrovich Pavlov (1849–1936) who was an extremely skillful experimental physiologist. His early work on mechanics of digestion and the alimentary tract led to a better understanding of gastro-intestinal function, but he is better remembered for detailed studies of the nervous relations of the tract. He developed the important concept of conditioned reflex action.

Many physiologists have devoted their attention to the problems of humans and lower mammals because they are relevant to medicine and the alleviation of disease. Because every disease is associated with a disorder of function, no physician can diagnose or treat patients intelligently without knowing about the physiology of cells, organs, and the organism as a whole. This is as true of disorders of the mind as it is of the body.

Scientists in many fields must have a sound knowledge of physiology. In so far as psychology explores mental or emotional processes, it deals with the physiology of the brain, endocrine organs, and other systems. Drugs alter the behaviour of some cells. Pharmacologists, therefore, must begin their studies with a knowledge of normal cell function and alterations. Reproduction is a physiological process, so even the social problems of population control are intertwined with physiological knowledge. Physicians who must guarantee the safety of man in extreme environments as in space or high altitude aircraft flight can do so with greater assurance because of the knowledge accumulated by physiologists about respiration, circulation, and the response of the human organism to gravitational forces and atmospheric variables.

Specialties

General physiology

How organisms as diverse as bacteria and man obtain food and convert it to energy, establish cell volume and composition, and produce electric potentials across cell membranes are all objects of investigation for general physiologists. Their strategy to discover a law that governs a process is to find organisms in which a certain process is highly developed and easily studied. The sea slug, *Aplysia*, has a simple nervous system made up of relatively few large cells. Our understanding of the function of the central nervous system of higher organisms has been enhanced by information obtained from electrophysiological studies on the nerve cells of these animals. The red blood cell has been used to study the transport of sugars and ions because it is easy to isolate and study, and its metabolic machinery is relatively simple. Such studies have shown how potassium is accumulated by cells in contrast to sodium and why cells do not swell or burst from the uptake of water.

Modern physiology is less concerned with processes and more concerned with regulation of processes. Both depend on cell membrane function, so many physiologists work in that area. They have created synthetic membranes which behave electrically like living ones. These should help us to understand how membranes protect the cell from the environment and facilitate nourishment. We still do not know how a membrane "perceives" light or how it produces the sensations of taste or smell. Perhaps a physiologist will have the distinction of providing the answers.

Mammalian physiology

Mammalian physiology deals with the function of organ systems of mammals. The problems it treats bear directly on human and veterinary medicine. Mammalian physiologists subspecialize. They work in a particular area, for example, studying the function of the reproductive system or nerves.

Mammalian physiologists interested in gastrointestinal function try to understand how animals ingest and digest foods and how this food is absorbed by the blood stream and distributed throughout the body. How this distributing system operates, with emphasis on the heart which pumps the blood, is the province of cardiovascular physiologists. Metabolic specialists are concerned with the use of food by the cells in the production of energy and its use. This speciality blends physiology and biochemistry. Renal physiologists investigate kidney function to learn how this organ rids blood of the waste products of metabolism while it conserves vital chemicals. The transport and exchange of gases between body fluids and the external environment is the realm of pulmonary physiology. Some specialists delve into the complex mechanisms that govern reproduction and the changes that go hand in hand with it, such as lactation.

Neurophysiology is one of the most challenging subdivisions of mammalian physiology because it deals with the most complicated of biological devices, the human brain and its extensions, the sense organs. This integrative system also controls, through the pituitary gland, the function of most of the endocrine glands. These glands, in turn, regulate growth, reproduction, metabolism, and a host of other processes.

Mammals are amazingly well-engineered units complete with sensors, feedback control mechanisms, data processing systems, and communication systems. The traditional tools of biology are not sufficient to probe the operation of these vastly complex controls. New and more sophisticated techniques are needed. Many of the answers are being sought and found by means of systems engineering and mathematical modelling and with the aid of biosensor devices and computers.

Comparative physiology

Comparative physiologists study the life processes and dysfunctions of different kinds of animals. They seek answers to questions such as: How can birds fly to great heights without suffering from lack of oxygen? How do polar bears survive in the arctic cold? How can lizards live in the desert? The scientist who wants to answer such questions must have, in addition to his knowledge of physiology and allied sciences, a special ability to plan experiments and to recognize the peculiar problems associated with many species.

It can be hard to find out what an animal is doing. When bats fly about in the dark, we now know they guide themselves by sonar, listening for reflection of sound to locate objects around them. Some insects preyed upon by bats use sonar frequencies which interfere with the bat's use of its own signalling system. Humans cannot hear any of these sounds, and they were discovered only through the use of special recording instruments.

The comparative physiologist must be imaginative, adaptive, and energetic. He may have to travel to tropical rain forests or mountain tops where particular animals live, or work on ships at sea. Conditions may be primitive. Far from the conveniences of the laboratory, the comparative physiologist may have to design his own equipment. This work demands constant ingenuity and enterprise.

Farm animals and pets have their own special physiologies, and understanding these is invaluable for rearing healthy animals. Comparative physiologists support veterinary medicine in much the same way that mammalian physiologists support the practice of physicians. The problems are of great interest in themselves and may also have immediate bearing on human life. Again and again, new ideas in comparative physiology lead to insights of great importance for understanding our own health.

Reproductive physiology

It is useful to begin with a definition. The biologist accepts as axiomatic that in nature the primary and essential task of the individual of the species is to assure survival of the species. More than that, it is essential to produce offspring which is in turn capable of producing normal healthy offspring. In most animal forms, certainly in higher forms, this process involves the participation of two partners: male and female.

The genital system is clearly prime in the reproductive function, but all of the various organs and organ systems of the body are involved in achieving the successful generation of offspring. Each biological event or sequence of events within the organism, male and female, does or may influence reproductive performance. Yet, reproduction involves more than the sum of the primary and secondary activities which are concerned in conception and the birth of a healthy offspring. Reproduction involves more than conception and birth. It must be viewed as extending beyond the limits of the physiology of reproduction, and if it is seen from the standpoint of ecology, reproductive capacity will be found to depend on demographic, social, cultural, economic and anthropological factors.

Environmental aspects: the cultural, social, anthropological factors, the economic and class difference, religious taboos, rituals, fertility rites; food habits and practices of hygiene, and the host of other features of geography and climate which have an impact on man's (and other animals') behaviour must be recognized. Thus, it must always be borne in mind, whether we study fertility in man or other animals, that fertility or infertility is a manifestation of an extremely diversified combination of biological, physical and social elements.

Thus the reproduction physiologist has a very challenging field for the exploration of not only the biological phenomena, but the vast array of social and environmental influences. The importance of understanding reproduction can be appreciated by the widespread existence of infertility, of barrenness of couples who desire children – and, on the other side, uncontrolled reproduction leading to global overpopulation.

Environmental physiology

The quality of our national environment is a matter of grave concern. By the late 1960's, the rapid rise in air and water pollution, use of pesticides, industrial noise, and depletion of natural resources led to the enactment in the U.S.A. of the National Environmental Policy Act. This Act declared "A national policy which will encourage productive and enjoyable harmony between man and his environment ..." The growing emphasis on preserving and improving the environment has provided an impetus for the relatively new discipline of environmental physiology. This is one of the most applied specialities and one which allows a physiologist to make a very direct and positive contribution to society.

An environmental physiologist studies the response of humans to those forces and stresses which occur in the normal environment. He might, for example, be interested in the effects of heat stress on people living and working in the tropics. Possibly even more important to the environmental physiologist is the study of human response to altered environmental conditions. Man has drastically changed the environment to meet his own needs with the tools, power, and technology that have become available since the Industrial Revolution. Travel in high-speed ground and air transport vehicles, for example, imposes unique stresses with which the environmental physiologist of today must deal.

Unfortunately, man-made modifications at times have had a damaging impact on human health and welfare. Air pollution and the introduction of dangerous chemicals into the waterways are two familiar examples. Now, as awareness of these side effects grows, much is being done to reverse the trend. Even so, as civilization advances, there will be a constant need to assess the physiological effects produced by change. Through the development of programmes and measurement procedures to determine the effect of new technology on humans, environmental physiologists will do much to insure an environment of high quality for generations of the future.

Preparation for a career

There are many challenging positions open to environmental physiologists in universities, government service, private research institutes, and in industry. These include administration of large-scale programmes, training of different occupational groups, development of new biomedical and protective equipment, and conducting basic research.

A knowledge of chemistry is very important to an understanding of physiology. Essential characteristics of biological structures and reactions are described in the terminology of organic chemistry. Analytical chemistry provides techniques which permit detection, identification, and quantitation of chemicals in the body. Physical chemistry provides a systematic framework for analyzing the flow of energy that takes place during such processes as muscular contraction. This discipline also provides techniques for analyzing the physical effects of interactions between molecules. For example, the secretion of lipids into the air sacs of the lungs is necessary to reduce the stiffness of the lungs sufficiently to permit lung inflation. Introductory chemistry will be useful to most laboratory assistants in physiology. On the other hand, even the most advanced courses may not provide the information essential for work in some areas. In that case, it may become necessary to conduct research in chemistry to develop the chemical knowledge needed to interpret a physiological study. This is where team work comes in.

Knowledge of physics is also basic to the understanding of physiology. Electrophysiology, hearing and speech physiology, and the physiology of vision are based on principles of electricity, sound and light. Studies of blood flow and air movement are complex extensions of hydrodynamics. Temperature regulation in animals and man is a complex example of the physics of heat flow. The use of radioactive tracers as a tool in physiology requires an understanding of radiation physics. All types of instruments have become important in the study of physiology. The physiologist must therefore understand the physical and engineering principles which govern and limit the use of these instruments if he is to avoid misinterpreting his experimental data.

Biology and physiology are obviously important for the training of a physiologist. The need to study mammalian structures as a basis for understanding the functions of mammals is more apparent than the need to study the whole range of the animal kingdom. Many physiological problems have been solved, however, after an investigator identifies some processes in a non-mammalian species. A squid nerve fibre, which is large enough to permit measurement of contents, has helped us learn many things about the function of a human nerve which may be only 1/1000th as large in diameter. The nearly transparent skin of the toe webbing of a frog has helped us to understand capillary circulation beneath the less transparent human skin. Both physiological technicians and physiologists need to know the normal biology of their experimental animals. However, they should also know about the biology of the parasites and bacteria which can alter the physiology of these animals.

The mathematics a physiologist needs is not a mathematics of number manipulation but a knowledge of principles. Machines and packaged analysis programmes are available to do the work of statistical analysis, but the physiologist must understand probability theory to select the proper programmes and interpret their results. He should understand calculus because it is the mathematics of change. He will turn to calculus to manipulate ideas about the changes which are basic to life. The dramatic developments in computer technology and the availability of highly sophisticated equipment at modest costs have made computers important to the physiologist for research and for teaching. Therefore, basic understanding of computer sciences will become progressively important to the student.

Manual dexterity has been an important factor in a number of advances in physiology. The dissection of single living cells and sampling of fluid from single kidney tubules are examples of skilled, mechanical manipulations which have made possible the acquisition of critical experimental data. Skill in manipulating small and delicate objects with and without instruments is needed in both experimental and academic physiology.

ECOLOGY

Profile No. 5

ECOLOGY

by

Dr Malcolm J. Coe
(Lecturer in Animal Ecology and Tutorial Fellow in Zoology,
St. Peter's College, Oxford)

In the 18th and 19th centuries, European naturalists travelled around the world collecting a vast variety of animals and plants from a wide variety of habitats that ranged from rain forests to hot deserts and cold polar regions. These collections provide a valuable record of the way in which living organisms are distributed and adapted to a whole series of biotic and non-living (sometimes called abiotic) variables. Indeed, it was as a result of such observations of natural history that Charles Darwin and Alfred Russell Wallace made the first public statement of their theory of evolution by natural selection at the Linnean Society of London in July, 1858.

The term ecology is variously attributed to Reiter (1865) and Haeckel (1869) who took the term from a combination of the Greek "Logos", meaning the study of, and "Oikos", or house. Thus we may think of ecology as being the study of the house of nature or the natural environment. More strictly, however, the term denotes the scientific study of the interactions that determine the distribution and abundance of living organisms (Krebs, 1972). Such studies encompass not only interactions between living organisms themselves, but more important those between them and their non-living surroundings.

A first awareness of ecology, in the sense that we understand it today, must be almost as old as man himself, for the first hominid hunter gatherers must have survived through their basic ecological knowledge. They must have been aware of the distribution of the animals and plants on which they fed, and they would have organized their day to day activities in a way which reflected seasonal changes in the abundance and availability of the organisms on which their survival depended.

It is difficult to describe ecology as a neat single unit of study, for by its very nature it is inter-disciplinary and every year the contributors to the science are increased as the questions we wish to answer become more detailed and complicated. Basically, we may begin with the study of evolution itself, for the study of adaptation is the starting point of virtually all ecological studies, while ecological genetics helps us to understand the mechanisms whereby such adaptive changes are brought about. At the same time, the study of plant and animal physiology is central to the ecologist's understanding of the way in which animals and plants feed, breed and respond to physicochemical factors of their environment. When the ecologist studies a single animal or plant species, the behaviour of that organism, in response to the non-living environment, on its own and other species has developed into another important area of study which lies at the interface between ecologists and those biologists who are predominantly concerned with behaviour.

Recently mathematics has also been employed widely as a valuable tool in attempts to construct models of predators and their prey, the competition for food/survival within one or between different ecosystems and for entire ecosystems. This development has introduced an important predictive element into ecology.

An ecologist considering non-living influences must study the effect of physicochemical factors of the environment on the observed pattern of a species' life history and population dynamics. In order to do this, we can divide the world into terrestrial and aquatic (fresh water and marine) environments. By the measurement of climate, we may learn that a particular insect's distribution is limited by the prevailing winter temperature, or a marine organism by salinity. The geological nature of a particular part of the earth's surface together with its climate will, to a large degree, determine the type of vegetation we expect to find there. This in turn will have a profound, and possibly even predominant, effect on the fauna of the area.

One of the best ways to understand what an ecologist does is to look at a number of different units of study, each of which represents an increasing degree of complexity. The first of these units must be the individual organism, which is the essential starting point of many behavioural or ecological investigations, for organisms interact both with one another and with the environment that surrounds them. A group of a single species in one place, at one time, represent what we call a population, and here the interactions will of necessity be more complicated than those of an individual organism. Except for laboratory studies, there are probably no environments in the world that are dominated by a single species to the exclusion of all others, so we must next look at the community which represents a whole range of populations of species interactions with each other and their environment both in space and time. At a level of even greater complexity we may study a biome. This is a characteristic broad type of ecosystem possessing similar communities. Thus, we might examine a temperate deciduous woodland, a tropical grassland or a rain forest, irrespective of the continent in which it is found. High levels of rainfall in the tropics may always be expected to produce something which we recognise as rain forest, even though its constituent parts may be quite different and various examples may have species structures which have been reached by quite different evolutionary paths.

The size of a unit which we choose to study will depend on our particular interests or the hypotheses that we have formed from observing living organisms in a particular locality. We might therefore choose to study the island of Aldabra, in the Indian Ocean, as a typical low lying coral atoll in a tropical sea. In so doing, we might attempt to study the whole ecosystem which comprises an assemblage of seabirds, land animals, vegetation and a variety of different habitats. Alternatively we might restrict ourselves to studying the creatures that occupy the vegetation of eroded coral limestone (Champignon), or even the population of giant tortoises that survive on this isolated land mass.

If we choose the last of these, we shall have to ask ourselves a series of basic questions about the species under study. First, we would investigate their numbers and distribution in the environment. This would lead us to ask what factors in their behaviour and surroundings (shelter, water, food availability) determine their observed distribution. If we collect our information carefully, we should discover that their distribution is determined largely by the distribution of their food plants. A study of the climate would indicate that variations in rainfall affect the amount of food available to tortoises and this, in turn, controls the number of eggs the female can lay, which will regulate their numbers. This last conclusion is ultimately central to virtually all ecological studies, for it attempts to discover what regulates the numbers of plants or animals which an environment can support.

This information is important in the study of a single species, for it allows us to look at a variety of parameters, all of which will interact to produce a balanced ecosystem in which the relative numbers of organisms remain remarkably constant from year to year unless some unpredictable event perturbs the system.

If, however, we are to understand these interactions and the various controls that maintain the balance of an ecosystem, it is necessary for us to look for a common denominator that either directly or indirectly affects all parts of that system. We have been aware for a long time that some insects eat plants, while shrews, in their turn, eat insects and owls eat shrews. Thus, we can place organisms in simple food chains or more complex patterns in which, by examining all the different organisms eaten by shrews, and all the different small mammals eaten by owls, we arrange them into a food web. This serves to indicate the effect of disturbing or removing a single element from the web, and the way in which the system will need to adjust itself to a new level of balance or stability. Ecologists usually distinguish two types of food web, the grazing food web in which organisms feed on living material, and the detritus food web where they feed on dead material.

If we examine any of these food webs it soon becomes apparent that each group of organisms feeding on another group is deriving energy and nutrients from these activities. The ultimate major source of energy for all ecosystems is radiant energy fixed by plants during the process of photosynthesis, while nutrients are obtained from the soil through the plant roots. We therefore call all these autotrophic organisms primary producers, since they construct complicated organic molecules from simple materials. Since animals are ultimately dependent on these producers we call them the consumers. We have already seen that in a food web or chain, herbivores eat plants and carnivores eat herbivores, so we may distinguish primary, secondary and tertiary consumers. Since each level of consumer will require energy for respiration and synthesis, only about 10% of the energy will be available for the next level of consumer. Thus in practice there are never more than five levels of consumer distinguishable in any ecosystem. Energy is therefore said to flow through an ecosystem according to the first and second laws of thermodynamics.

Although radiant energy may be considered to be a potentially infinite source of energy which continuously generates new primary production, nutrients are a finite resource which cycle within the ecosystem. Thus, when dead plants or dead animals and their waste products are returned to the environment, they are broken down by decomposers (bacteria and fungi) and stored in what we call the nutrient pool, and can be used again by primary producers. It is the rate at which these materials are made available that will, together with the local climate and soil conditions, determine the rate of plant growth and hence the biomass of animals that a particular environment can support. These concepts not only comprise the basic foundation of ecological science but also represent a framework within which all human exploitive activities can and must be viewed.

Attempts at modelling natural ecosystems may seem to represent academic activities which are rather remote from the needs of the real world. The problems of providing food for an ever increasing human population make it clear that if man is to survive on this planet, much time and energy will have to be devoted to devising production and land management practices based on sound ecological principles. There already exists much of the essential practical and theoretical framework which will in future be far more widely used in devising programmes of rational exploitation, even if some of the ideas do not quite fit modern economic theory. Recent increased awareness of the dangers of pollution and related human influences on biogeochemical processes has led to the assumption that these activities are solely related to "ecology" or "environment". In reality, although the effects are ecological, to a large degree, the solution of such problems lies with the industrialists, developers and politicians. There is still, however, a great gap between the standard of living of the developed and developing world and it is in closing this gulf that the ecologist will undoubtedly play an increasing role, whether it be in the fields of conservation, land management, tourism, amenity or the broad exploitation of natural resources.

Ecology began with the study of natural history and this is the level at which many young people are drawn towards this study as a career or merely as a basic philosophy for living. There are today few fields of animal biology that do not provide a place for the ecologist, whether in teaching, pest control, marine biology, fisheries, parasitology or wildlife management. Since, however, ecology is totally interdisciplinary in its content and outlook, it is essential that those contemplating being trained in it should make themselves aware of the whole of biological science and its remotest constituent parts in order that they may give this subject the breadth of understanding that it demands.

Further reading

Owen, D.F. (1980) *What is Ecology?* (2nd edition). O.U.P. 188 pp.
Phillipson, J. (1967) *Ecological Energetics*. Edward Arnold. 51 pp.
Ricklefs, R.E. (1973) *Ecology*. Nelson. 861 pp.
Whittaker, R.H. (1975) *Communities and Ecosystems*. Collier Macmillan. 385 pp.

ETHOLOGY

Profile No. 6

ETHOLOGY

by

Dr Peter J.B. Slater
(Lecturer in the School of Biological Sciences, University of Sussex)

Ethology is the branch of biology which is concerned with the behaviour of animals. It is a subject akin to psychology, but its emphasis is rather different, for while ethologists are interested in the natural behaviour of animals, and how evolution has adapted it to their surroundings, psychologists tend to be more concerned with animals' capabilities, such as the tasks that they can be trained to master. It is sometimes said that the psychologist puts an animal in a small box in his laboratory and looks in to see what it is doing, while the ethologist goes outside, puts himself in a box to avoid disturbance and peers out to watch how the animals around him behave. The popular image of the ethologist is certainly of someone who, with boots, anorak and binoculars, spends most of his working life in the open air, observing and describing in great detail the behaviour of the animals which he is studying. It is an image which applies to some of them, but many ethologists have found that the answers to their questions can be found only by carefully controlled experiments, which are more easily carried out in the laboratory than in the wild.

Although people have been interested in the behaviour of animals for a long time, and Darwin, for example, wrote quite a lot about it, the real growth of ethology started with the work of Konrad Lorenz and Niko Tinbergen in the 1930s. Anyone who has read Lorenz's delightful book *King Solomon's Ring* will remember the amusing scene in which he was dragging himself, hunched up, through some long grass, making the odd quacking sound, when he looked up to see a line of horrified faces staring at him over the garden fence. What the onlookers could not see was a line of small ducklings following him through the grass, but I doubt if they would have been any less shocked had they done so! Fortunately, Lorenz was not too worried about what people thought of him, for what he was up to was science, not just fun and games. He had discovered what is called imprinting: that young birds, when they hatch, do not know what their mother looks like, but follow the first moving object that they see, learn about it and become attached to it. In the wild this usually works well, for the object will be their mother and so they keep close to her and imprint upon her. But experiments have shown that they can become inseparably attached to anything from a distinguished professor to a red watering can if they are shown one at the right time. In this, and many other aspects of behaviour development, the timing is crucial; once the chicks are a week or two old it is too late and an object which they would have followed earlier will simply terrify them if they have not seen it before.

Much of the behaviour of the insects, fishes and birds which early ethologists studied in detail is deeply rooted, and each species has an array of what are called "fixed action patterns", many of which vary very little from one animal to another. From this it was at first thought that such patterns were innate or genetically determined, but it was soon realised that this was an oversimplification. Many patterns, like the song of a bird or the attachment of a young animal to its mother, have been found by experiments to depend on learning; they develop wrongly when the animal is not in a natural situation. A particular fixed action pattern may always appear in the wild because the animal develops in the right environment, but behaviour patterns may turn out to be very much less fixed when animals are reared in environments like the laboratory to which evolution has not adapted them. Imprinting is a good example of this variation. As we have seen, animals can become attached to any one of a wide variety of different looking objects. With sounds, however, they are much more fussy: ducklings prefer objects which quack and chickens those which cluck. Ethologists have discovered many intricate ways in which behaviour patterns become formed as young animals grow up. The usual process is a complicated interaction between the animal and its environment; it hardly does justice to this to call some patterns "innate" and others "learnt".

The study of development is just one area in which ethologists have made great strides by means of structured experiments instead of simply observing how animals behave. Some experiments can be carried out in the wild but this is not always easy; results can be hard to interpret when pouring rain, a nearby owl or a passing Concorde are factors which have to be taken into account. To experiment on an animal's development one must place careful restrictions on the experiences to which it is subjected and rearing in captivity is the best way of doing this.

Examining the causes of behaviour, and the mechanisms underlying them, is another area of active interest in ethology which has benefited from laboratory experiments. The aim here is to discover why animals do what they do when they do it. One aspect of this is how animals react to different stimuli in the outside world, and experiments both in the field and the laboratory can examine how they respond to such things as models of predators or tape-recordings of alarm calls. But the way in which they behave depends not only on the outside world; it also varies from time to time according to changes going on within the animals themselves, and these are loosely labelled as shifts in motivation. Changes of this sort are also of interest to psychologists, and the two subjects have much in common in this area. For example, one of the most effective ways to study when an animal feeds and when it drinks is to place it in a box with two levers, one of which can be pressed to deliver food and the other water. Psychologists and ethologists have both used this method. Another overlap is with physiology. The tendency of animals to court or to fight depends in part on their hormone levels, and these can be raised or lowered experimentally to reveal their influence on behaviour. Ethologists have also begun to examine the way in which the brain affects behaviour by noting the effects on the various regions by either removing or stimulating them. But, physiology aside, it is the study of the causes of behaviour which requires ethology to become most technically sophisticated, with observation on closed-circuit television and computer analysis now used as fairly standard procedures.

The study of ethology stands also on the borders of ecology and evolutionary biology. Being trained in biology, ethologists have always been interested in how behaviour evolved and in the ways in which it is adapted to give an animal the best chances of surviving and reproducing. Recent advances in the theory of evolution have strong implications for behaviour, such as the suggestion that animals always behave selfishly except where this conflicts with the interests of their genes, as when they have dealings with relatives. By studying the behaviour of animals in social groups in the wild, ethologists can test such theories with their own observation. Sociobiology, as this field is often called, has become something of a growth industry. Studies of social behaviour like these examine how animals behave towards each other; more generally, behavioural ecology is concerned with the way in which their behaviour is adapted to all factors of their environment. Many species have now been studied in the field and, as a result, it has become possible to relate differences in their behaviour to differences in their surroundings. To take an example, there are wide variations between species of monkey in their food, the size of the troops in which they live, the sex ratio in the troops and the area over which a troop will range. Studies of these differences have suggested that large home ranges and troop sizes have arisen as an adaptation to an unpredictable food supply, it being necessary to have a large range to ensure that some is always available. This conclusion is based on comparisons between species; another approach is to compare the actual behaviour of one species with a theory of how it should behave if it were well fitted to its environment. For instance, studies of foraging have revealed how animals look for food when it is distributed patchily in the area over which they are searching, as so often in nature. The pattern of their search can be compared with other possibilities to see whether they use the best possible strategy and to discover their rules for finding food.

Ethology, then, is concerned with several different types of question. How does behaviour develop? What are the internal and external factors which affect it? How has it changed during evolution? To what extent does the behaviour of an animal adapt it to its environment? Some ethologists are mainly interested in one or other of these, but for others the fascination lies in how they all fit together. Most ethology is, however, pure science, aimed at discovering the principles underlying animal behaviour and without obvious applications. This is not always true: studies of rare or endangered species are important for conservation, and those of the behaviour of farm animals have implications for food production and welfare. Animals can also be used to throw light on problems of human behaviour; processes can be studied at a simpler level, and it can then be considered whether they operate similarly in man. Into this category comes work on how the responses of animals are affected by various drugs, or on the long term effects of separating an infant monkey briefly from its mother. The methods of detailed observation and analysis, developed by ethologists for studying animals, have also been found useful in some studies of human behaviour.

Popular books on ethology, which are usually based more on speculation than observation, often suggest that its relevance to man is great because we are animals and our behaviour is the product of evolution. But on this point most ethologists are cautious, for they know just how much the behaviour of animals can differ, even within closely related species, and they also know that our biological past lies a long way behind us. On the other hand, even if one cannot generalise simply from animals to man, ethology provides a marvellous source of ideas about behaviour which we can think about, discuss and test on humans. Although it is a comparatively small and recently developed branch of science it has already had a considerable impact on the way in which we view ourselves and the world around us. All the indications are that ethology will continue to flourish, giving us insights into the organisation and evolution of behaviour, knowledge of the mechanisms whereby it is produced and, perhaps most fascinating of all, more beautiful examples of the subtle ways in which animals are adapted to the world in which they live.

Suggested reading

Dawkins, R. (1976) *The Selfish Gene*. Oxford University Press.
Lorenz, K.Z. (1952) *King Solomon's Ring*. Methuen.
Manning, A. (1979) *An Introduction to Animal Behaviour*. Edward Arnold.

PSYCHOLOGY

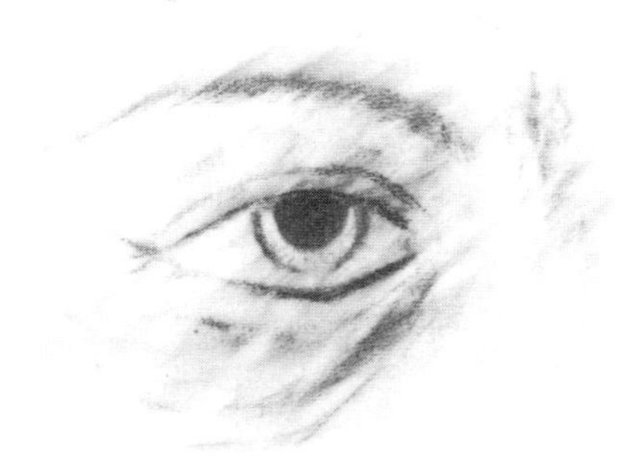

Profile No. 7

PSYCHOLOGY

by

Professor H. Gwynne Jones
(Professor of Psychology in the University of Leeds)

Psychology extends the biological study of structure and function to the realms of behaviour and experience. Psychologists' interest in behaviour is shared with zoologically trained ethologists. At one time, whereas ethologists studied animal behaviour in the natural environment and for its own sake, psychologists studied animals' reactions to artificial experimental laboratory situations mainly to elaborate and test theories of human learning and problem solving. More recently, ethologists have begun to apply their techniques to the study of human interactions, especially those between mother and child, and psychologists, recognising the artificiality of laboratory experiments, have undertaken animal field studies. The study of human experience, our conscious thinking and feeling, although not divorced from biological ways of thinking, has its main historical roots more in philosophy and there are still active controversies as to how this study should be approached. The social sciences, sociology and anthropology are also relevant to psychology. All behaviour has a biological basis but is greatly affected by the personal and social context in which it occurs.

A person's behaviour is characterized by its flexibility in that it is continuously adjusted according to the incoming flow of information from the environment which is interpreted and selected according to the person's understanding of the situation to maintain a path towards his objective, be it long term or short term. "Feed-back" information about the effects of his actions is of prime importance to his control and progress. Regarded in this way a human being is a complex but integrated self-regulating organism in intimate ecological interaction with its environment. This description would also apply to sub-human species but the greater complexity of man's brain processes and particularly his use of language transforms the possibilities. Language as a symbolic system permits the elaboration of abstract intellectual concepts and as a means of communication has fostered the development of intricate social organisations and has greatly enhanced the importance of social aspects of the human environment. The resultant complexity of the human organism requires analysis at at least three interrelated and interacting levels. One, shared with all animals, is the level of biological functioning to satisfy needs such as hunger and thirst. Another is the level of social functioning according to the rules and roles of the society and culture to which the individual belongs. The interplay between these two is illustrated by the customs and taboos related to, say, eating or sexual behaviour. There is also what may be described as a humanistic level organised around a person's values, beliefs and sense of a unique identity or "self" exercising free-will. All levels reflect the complex biological nature of man and are the products of evolution. The problems of dealing with them within one framework of study makes psychology a challenging subject which has to bridge the gap between the biological and social sciences without neglecting its earlier roots in philosophy from which it has inherited some intractable problems.

Experimental psychology can be regarded as examining the manner and establishing the general laws by which information is processed. Thus *perception* is the psychological function concerned with the organisation and coding of the environmental stimuli impinging on the sense organs. The psychologist is interested in matters closely related to sensory physiology such as theories of colour vision but, more typically, is concerned with our direct experience of an external world of objects in space, near or far, still or in motion. In general, we perceive the world with much greater accuracy than expectations based on physics or physiology would allow. For example, objects are perceived as constant in size over a wide range of distance despite vastly different sizes of images produced on the eye's retina. At times, however, perception distorts reality as illustrated by the popular illusions often portrayed in magazines. Much research is concerned with the question of whether we learn to see the world as we do or whether this depends on inborn characteristics. Both factors have been shown to be important and an individual's motives at any particular time may also affect his perceptions.

Learning, the ability to store past experience and to bring it to bear on present events, has been the major pre-occupation of many psychologists. Conditioned reflex and various forms of trial and error learning have been studied in detail in animals such as rats and pigeons. Given appropriate training under the guidance of selective reward these animals are capable of very complex learned behaviour to which the "cognitive" qualities of insight and understanding more commonly associated with human learning are not necessarily irrelevant. Similarly the more mechanical forms of learning play a part in complex human learning and memory, especially those involving emotions and motor skills. Experiments with primate species such as chimpanzees support the concept of continuing evolution of man and other animals in this as in other functions. Man is particularly gifted in his ability to transfer learning from one task to another and to acquire skills vicariously, especially in the social sphere, by imitating the behaviour of others. However, man's supreme advantage in learning is his language, which not only facilitates direct learning but provides access through books to the knowledge and experience of an entire culture.

Learning and language are involved in all types of human *thinking*, *problem solving* and other *cognitive* processes. We can abstract the common properties of a group of objects, events or relationships to form a named concept as a unit for further learning and thought. Rules refer to the relationships between such concepts and, by a process of verbal *reasoning*, new rules can be deduced from two or more others, as is well illustrated in Euclidean geometry. Problem solving is essentially the creative application of established skills and rules to new situations. Non-verbal codes such as the numerical and spatial codes of mathematics are also important in these areas.

Language itself has become a specialist subject of study for psychologists. *Psycholinguistics* is concerned with both the symbolic and communicative functions of language and especially with the manner in which a child masters the enormous task of learning to speak its mother tongue, a task quite different from the formal later learning of a second language. Recently, great interest has been aroused by language-like systems successfully taught to chimpanzees. The question of whether these are truly akin to human language is now a matter of keen controversy.

Behaviour is actively directed towards discernible ends. Concepts of *motivation* are concerned with these energising and directive aspects of behaviour. Like ethologists, psychologists make use of theories of instinct, but, unlike them, have stressed the importance of the physiological controls which together maintain the body in a state of equilibrium and underlie the universal drives of hunger, thirst, sex and the avoidance of pain as well as incentive value of environmental stimuli relevant to these drives. Other social and humanistic motives have already been mentioned but there also appear to be more general energising motives related to tendencies to seek stimulation and to explore and acquire mastery ot novel situations.

All these psychological functions undergo a changing course through the life-history of an individual and *developmental psychology* has become a very active specialist area of study and research. Until comparatively recently a new-born infant was considered a helpless, passive receiver of attention. Improved methods of study have now demonstrated that, even in the first days of life, a child has a growing and reactive awareness of its world and even exercises some measure of control over it. Later maturational sequences of motor and cognitive development have been studied in detail and a great deal of research is concerned with the interaction between biological maturation and learning. Adolescence with its problems of emancipation from parental control and search for an independent identity has been much studied but there is now a growing interest in the nature and problems of maturity and old age.

Social psychology is another field which impinges on all others but has a special concern with the manner in which people interact with each other, in pairs and in groups, small or large. One topic of active interest is interpersonal perception, the manner in which we see, judge and are attracted or repelled by others. An important and often distorting factor is the operation of *stereotypes*, a group of characteristics thought to be true of a class of people without supporting evidence. Negative attitudes based on such stereotypes underlie racial and other forms of *prejudice*. Many social psychologists are therefore engaged in studies of how attitudes may be changed.

Abnormal psychology is concerned with the maladjusted behaviour and disturbed emotions of those suffering from various types of mental disorder. Emotions such as fear, anxiety, anger, joy are excited states closely related to motives. They involve physiological arousal, especially of the autonomic nervous system, a pleasant or unpleasant subjective experience and related behaviour. Some abnormal mental states reflect medical conditions, especially of the nervous system and are mainly the province of psychiatry, a specialist branch of medicine. Others are distortions of normal emotional reactions. Whereas positive emotions are associated with secure control of situations and the achievement of desired goals, negative emotions such as anxiety arise from precarious control, conflict and threat. Such stresses evoke in different people different degrees of distress and can lead in some to severe forms of neurotic disorder.

One field of psychology is devoted to the study of such *individual differences*, the manner of expression of the various functions described earlier which gives to a person a unique and recognisable *personality*. A large number of psychological tests and other assessment techniques have been devised to assess and describe an individual's style of social behaviour, his temperamental characteristics and his intellectual qualities.

Psychology is practised as a profession in a variety of its applications. *Clinical* psychologists work mainly in the Health Services applying their skills to the assessment and treatment of mental disorders. *Educational* psychologists are concerned with the educational difficulties and more general counselling of school children. *Occupational* psychologists are involved in the selection and training of workers for specialised tasks and in the "ergonomic" design of manufactured articles to improve the efficiency and comfort of their human use. For example, the instrument panel of an aircraft needs to display a variety of types of complex information in a form which is readily assimilated by the pilot from a brief visual scan. Other psychologists work in the Prison Service, the armed forces and various civil service contexts. All these vocations require some form of postgraduate training but the basic principles of applied psychology are included in most undergraduate courses.

Psychology may be studied at a University either as a single "honours" subject or as one of two main subjects for a "combined studies" degree. If it is the latter, the other subject is usually biological or mathematical in a science faculty, or sociology or philosophy in an arts or social science faculty. All psychology students need to acquire at least a conceptual appreciation of the functional anatomy and biochemical dynamics of the brain and nervous system. The experimental laboratory work and the more naturalistic practical field studies require competence in the statistical analysis of data. Nowadays, too, there is increasing reliance on computer concepts and methods, and psychology degree courses provide training and practical experience in these areas.

Not all psychology graduates ultimately become professional psychologists. Many work in fields such as social work, education or consumer research to which psychology is clearly relevant, but a substantial number take up vocations for which many other formal educational qualifications would be equally acceptable. However, the breadth of psychology is so great, dealing as it does with matters of human importance and providing a disciplined analysis of these in a manner which bridges the "two cultures" of the humanities and the sciences, that an undergraduate course in psychology provides an excellent general education in and of itself.

PARASITOLOGY

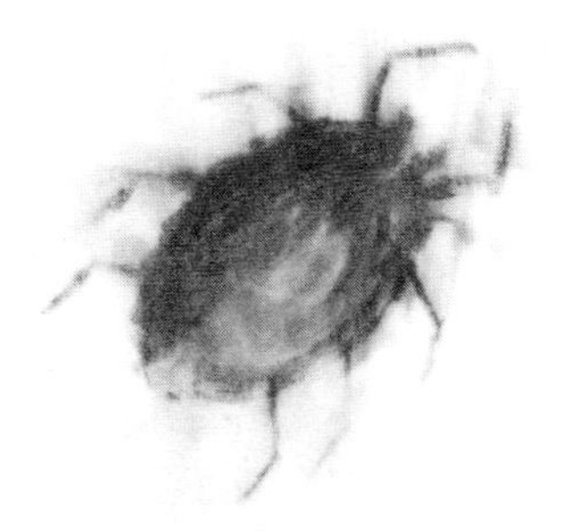

Profile No. 8

PARASITOLOGY

by

Professor Francis E.G. Cox
(Professor of Zoology in the University of London)

What is parasitology?

In parasitology, as in many areas of biology, definitions tend to be arbitrary but any organism that lives in or on another living organism can be considered to be a parasite. There are parasitic members in all groups of organisms ranging from bacteria to plants like mistletoe but most parasitologists are concerned only with animal parasites that live in other animals, including man. The very thought of a parasite like a tapeworm living in the gut frequently arouses mixed feelings of revulsion and fascination and, for hundreds of years, such animals have been sought out and studied wherever biology has been taught. Anyone who has studied biology will have heard of several parasites and seen the massive numbers that can be found in commonly dissected animals such as frogs and earthworms. The ubiquitous nature of parasites has made them the subject of many different kinds of investigation but their main claim to fame is as the causative agents of some of the most serious diseases of man and his domesticated animals.

Parasites and disease

All parasites cause their hosts some harm but this is seldom noticed unless the host is man or one of the animals he uses. In fact, parasites cause no fewer than five of the six diseases the World Health Organisation has selected as being of particular importance in the tropics and hundreds of millions of people are affected. The most serious disease is malaria, caused by a single-celled, or protozoan, parasite that lives inside the red blood cells of man. Over 200 million people are affected and in Africa alone over a million children die from the disease each year. Trypanosomes are also protozoans that live in the blood, in this case among the cells. In Africa, they are carried from man to man by tsetse flies and cause sleeping sickness. Similar parasites infect cattle and other domesticated animals, causing wasting and death and making it impossible to keep stock in over ten million square kilometres of Africa. Another trypanosome affects 12 million people in South America. A third protozoan that lives in the skin or internal organs causes leishmaniasis in various parts of the tropics and subtropics; in its worst form it is invariably fatal and at best it causes permanent disfigurement. Apart from the protozoa, most other parasites are worms and the most important worm disease in man is schistosomiasis. The worms live in the blood vessels of the infected person and lay eggs that pass into water to infect snails which are the other hosts of this parasite. Not all the eggs escape, however, and some lodge in various parts of the infected person causing massive local reactions that are always debilitating and often fatal. Another worm disease, filariasis, exists in several forms causing a variety of symptoms, including elephantiasis and blindness, and affects 300 million people.

As well as these important diseases there are parasitic infections that cause diarrhoea, wasting, blindness, deafness, damage to unborn children and have many other effects in both tropical and temperate parts of the world. Few animals are free from parasites and the loss of meat due to liver flukes and roundworms in the gut is astronomical. Parasites are particularly important in conditions of intensive rearing where epidemics cause the loss of millions of pounds worth of chickens and cattle each year. Even the home is not safe, for dogs and cats are frequently infected and may pass their parasites on to children or pregnant women, with tragic consequences.

Why is it then that, if we know so much about parasites, we do nothing to control them? The reason is both simple and complex. Parasites are very highly evolved and have found numerous ways to survive and although a lot is known about how they do this, it has not yet been possible to exploit the defects that have been discovered. Take vaccines for example. It should be possible to vaccinate men and animals against parasites but parasites have evolved numerous ways of evading the natural immune responses of their hosts, let alone artificial ones. Trypanosomes keep changing their surface coats so the parasite is always one step ahead of the immune response. Schistosomes absorb red blood cell substances and are therefore effectively disguised as "host". Other parasites remain within cells where they are safe or even live in phagocytic cells that should destroy them. The ways in which parasites evade immune responses are interesting from both theoretical and practical viewpoints and this subject is attracting the attention of some of the world's best immunologists. The possibility of using drugs presents even greater problems for, unlike bacteria, the cells of parasites are very like our own so it is difficult to devise a drug that is both effective and safe. The situation is so bad that virtually no new drugs have become available for any of the major tropical diseases for over 30 years and we are still treating malaria with quinine, which has been used for centuries, and chloroquine, which is nearly 50 years old. A major problem is that no sooner is a drug produced than the parasites become resistant to it. Malaria in many parts of the world is resistant to one or more of the commonly used drugs and this resistance sometimes spreads to related compounds.

Another method of controlling parasites is by attacking their vectors. Vectors are animals that carry parasites from one host to another and, apart from schistosomiasis where the vector is a snail, all the vectors of the major parasitic diseases are insects or other arthropods. The problem with insecticides is that they are difficult to spray in the parts of the world where they are needed, create environmental problems and also frequently and rapidly induce insecticide resistance. The twin problems of mosquitoes resistant to insecticides and malaria parasites resistant to drugs are common in some parts of the world. There are many alternatives to insecticides for the control of insects but these have not been effective and many parasitologists believe that vector control will be achieved only with new insecticides which will inevitably be costly.

It would be wrong to imply that nothing can be done to control parasitic diseases for there have been some notable successes, particularly against intestinal worms of domesticated animals. There are vaccines against lungworm and other worms in sheep and cattle and effective drugs against the common worms of dogs and cats. The real problems are presented by the many different parasites of man and the protozoan parasites of animals coupled with the difficulties in treating so many individuals scattered all over the tropics even with the drugs that we now possess.

The biology of parasites

Of course, not all parasites affect man and domesticated animals, and some do practically no harm at all. To ecologists, however, all parasites are interesting for they form integral and important parts of all ecosystems. To survive, parasites have to find and enter new hosts and also leave these hosts at a later stage in their life-cycle. The adaptations that allow parasites to enter food chains are as varied as any that exist elsewhere in the animal kingdom. Some use several hosts, an invertebrate, for example, a predator and a final carnivore as in the broad tapeworm of man which uses a crustacean, a fish and finally man. Others have much simpler life-cycles and the host may become infected by chance contact, by actively seeking out and eating the parasite or by being penetrated by it. There are many other kinds of life-cycle but most are characterised by the existence of both free-living and parasitic forms, and their ways of life are often quite different. In the life-cycle of the liver fluke, the adult worm lives in the liver and eggs pass into the intestine via the bile duct and so to the outside world. A freeliving stage emerges from the egg and bores into a snail where a massive phase of reproduction occurs. Eventually another freeliving stage leaves the snail and swims to a grass blade where it surrounds itself with a protective capsule and is eventually eaten by a sheep or another animal. Even now the life-cycle is not complete for the young fluke has to bore through the gut wall to reach the liver where it matures. These successive and quite distinct stages have different environmental problems to solve and the ways in which they manage are of considerable interest to physiologists and biochemists.

Even the simplest protozoa have complex life-cycles. Trypanosomes, for example, when residing in the blood stream of man, have available plenty of glucose which they respire to pyruvate which is excreted by the host. Suddenly they may find themselves in the gut of a tsetse fly which is not only much cooler but also fails to provide a ready source of glucose. The trypanosome solves the problem by metabolising a quite different group of substances, the amino acids, in the tsetse fly gut.

The myriad of parasite life-cycles attract the attention of those who like the detective work involved in actually discovering a cycle and also those who are interested in the ecological, physiological and biochemical implications of the various stages. There are certainly many more life-cycles to be traced. A fine example of the importance of this kind of work comes from studies on an ubiquitous parasite, *Toxoplasma*, found in virtually all warm-blooded animals including man, in which it may cause blindness and damage to the foetus. Although this parasite has been known since 1909 it was less than ten years ago that it was discovered to be merely one stage in the life-cycle of another common parasite of cats and since then over a dozen similar life-cycles have been discovered. Even in the malaria parasite some details of the life cycle are still not clear. The elucidation of the life-cycles is the first stage in the study of any parasite in the wild and as more and more are traced so will the foundations be laid for further ecological, physiological, biochemical and immunological investigations.

Parasitological problems for everyone

The study of the major parasitic diseases is attracting increasing attention as parasitologists realise how massive the problems are and how important will be their solutions. At the same time, the World Health Organization has launched a major campaign to find tools and scientists in various disciplines to help to solve what seem to be intractable problems. A similar interest is being shown by agricultural organizations. Parasitology, then, offers opportunities for scientists to participate in demanding academic challenges against a background of social need. There will be a continual requirement for specialists in parasitology, tropical medicine, biochemistry, physiology, pharmacology, immunology, ecology, and entomology as well as physicists, chemists and engineers, who will have to be responsible for designing the equipment to implement any control programmes, and statisticians and mathematicians to plan and evaluate them. The sociologists and economists are not forgotten for they will have to be involved in the changes to people's habits that control programmes will bring and, should these be successful, the growth of populations that will follow.

Further reading

Lyons, K.M. (1978) *The Biology of Helminth Parasites*. Studies in Biology No. 102. Edward Arnold, London.

Whitfield, P.J. (1979) *The Biology of Parasitism*. Edward Arnold, London.

Wilson, R.A. (1979) *An Introduction to Parasitology*, 2nd edition. Studies in Biology No. 4. Edward Arnold, London.

Wood, C. (ed.) (1978) *Tropical Medicine from Romance to Reality*. Academic Press, London.

PHARMACY

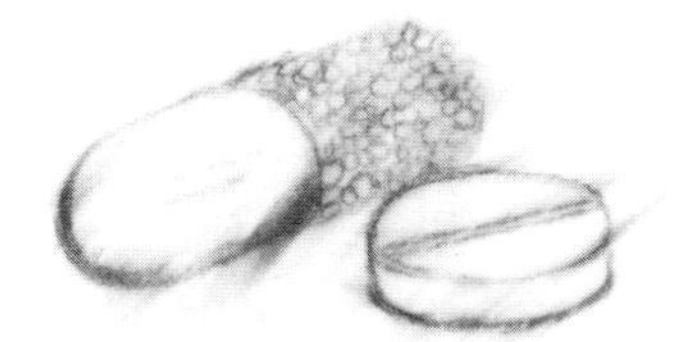

Profile No. 9

PHARMACY

by

Sir Frank Hartley
(Formerly Dean of the School of Pharmacy, University of London)

Pharmacy is a profession which is concerned with the preparation and supply of medicines either directly to the public or in response to the prescriptions which medical or dental practitioners have provided for their patients. In the U.K. the retail supply of medical products is restricted by law to Pharmaceutical Chemists who are registered by the Pharmaceutical Society.

Anyone seeking registration must possess a degree in pharmacy and must also have gained twelve months' experience in the practice of pharmacy under approved conditions. A degree course in pharmacy includes both practical work and the study of the chemical, physical and biological properties of the natural and synthetic substances which are used in medicines. A sound basic knowledge is needed of chemistry, biology, physics and some mathematics.

When a doctor or a dentist prescribes for a patient, he does so with particular intentions, and it is the duty of the pharmacist to ensure that the quality and safety of the preparations which he supplies will meet the prescriber's expectations. The pharmacist must therefore be concerned with the standard of the ingredients of medicines, and with the action, use and stability of the products which he supplies. The ingredients may be derived from natural sources such as plants or animals, and a pharmaceutical student must learn the principles of proper identification of animal and vegetable sources of drugs, their extraction, purification and characterisation. Both for the control of quality and the formulation of products for medical use, he needs understanding of relevant chemical, physical and physico-chemical principles.

The biological effects of substances may become apparent at different rates and may differ according to the manner and frequency in which the substances are administered to the patient. They may be given by mouth, by injection into body tissues or by absorption from the skin. The way a substance is absorbed and distributed in the body and its metabolism may influence both its activity and the duration of its effect. The effect of a drug may be influenced by yet other factors, particularly those involving the circumstances of the patient. Such factors include diet, exercise, the nature of the disease being treated and whether other substances or medicines are also being administered. A patient is provided with a substance in a formulation planned for his convenience. To design and prepare effective medical products the student must learn about the factors which influence the way in which these substances will act.

Chemical and biochemical reactions between ingredients obviously modify the properties of any mixture of them. Chromatography, spectrometry and titrimetry can be used in analytical studies which will determine purity and stability. It is also necessary to detect impurities, and determine their amount. Impurities may have been present from the start, or may have arisen from decomposition of the products in storage as a result of conditions of temperature, humidity and contamination by micro-organisms. The physical properties of substances may affect their absorption and distribution in the body tissues. Thus it is important to control the particle size of substances in suspensions of solids in liquids, or in emulsions.

The pharmacist is responsible for designing and supplying products which will have the expected effect, and he is therefore an important member of the health care team both in the prevention and the cure of illness. In earlier times, like the apothecary and the physician of old, he was concerned with the preparation and supply of products which relieved symptoms rather than cured ill health or disease. Now, however, he is concerned with substances which are highly active and which are capable of combating the causes of ill health, but which may also produce harmful or undesired effects in either normal or diseased bodies.

In former times it was the art of the apothecary, chemist and druggist to compound pills and potions which were acceptable to the patient in appearance and taste, and were presented in an elegant manner. All this demanded considerable manipulative skills, but they have now given way to the application of scientific knowledge to ensure that the treatment is what the diagnostician and prescriber had intended. The pharmacist of the present day must be concerned with the action and use of medical products in addition to having a knowledge of the chemistry of medical substances, and pharmaceutical aspects of their formulation.

The basis of pharmacology is the study of the action of substances on tissues, organs or the whole animal; clinical pharmacology deals with this as applied to humans, particularly in ill health. The pharmaceutical student therefore studies pharmacology to gain a good understanding of the mechanism of action of the substances which are used in medicine or in the preparation of medicines. This helps him to recognise the therapeutic incompatibilities which are so often the cause of illnesses resulting from medical treatment.

Study of the action of foreign substances on the body and its tissues requires an understanding of the normal functioning of the body. So the student must begin by learning about normal health in its biochemical and physiological aspects, and the ways in which it is disturbed by disease. He can then seek to understand the action of substances on abnormal or diseased states, and what is equally important, the effect on the substances themselves of being administered to man. If the substances used to combat disease do not retain in use the powers which laboratory study have attributed to them, no formulation made from them will achieve what the prescriber and patient had hoped. In other words the student must be concerned not only with the effect of the drug on man, but also of man on the drug.

Legislation during the past half century has established control over a schedule of so-called therapeutic substances. Wherever possible the purity and potency of such substances is defined in chemical terms but in some cases this is not possible and resort is made to a bioassay. A pharmaceutical student must understand the principles of bioassay, which will include statistical methods applied both to the design of the biological or microbiological experiments used for the tests, and for the interpretation of the results. The products which are prepared and standardised by such bioassay methods include vaccines (bacterial or viral), antisera, naturally occurring hormones such as insulin and very many antibiotics consisting of more than one molecular species.

A pharmaceutical student therefore needs to understand the principles and practice of bacteriology. He is less concerned with its clinical applications than with its part in the preparation of immunologically active products, with the contamination of medicinal products introduced into the body and with the biochemical changes which can be brought about by micro-organisms. If products are to be safe to use and able to be kept without spoilage, all bacterial contamination must be avoided. This requires the understanding and practice of aseptic conditions of manipulation, of preservation against growth of micro-organisms and of sterilisation processes. It is important to detect any inefficiency in such operations which may be revealed by the ability of a substance to cause a rise in body temperature or an immune response when administered to a rabbit.

Pharmacists are also responsible for the quality and standards of surgical ligatures and sutures and of surgical dressings; they therefore need to understand the nature, properties and performance of both natural and man-made fibres and materials. Pharmacy today thus extends far beyond the scope of the apothecary of old, or the chemists and druggists who founded the Pharmaceutical Society of Great Britain in 1841. The great majority of pharmacists are those who enter general practice. This means that they work in establishments which exist for the retail sale of medicines and health care products.

Many pharmacists are in professional full time employment in hospitals. The range of their activities includes compounding and supplying the medicine prescribed, the manufacture of sterile products, control of the quality of materials and products, and organising the collection, supply and retrieval of information on medical and pharmaceutical products to the medical, nursing and pharmaceutical staff.

Another group of pharmacists is employed in industry. They may work in research and development in chemical, pharmacological or pharmaceutical departments, or in technical services which include analytical and quality development and control. Their work may lie in production which would involve association with engineering design, operation and management, or they may be concerned with medical information and responsibility for registering products.

A pharmacist has professional responsibilities in his dealings with the general public and other professions and also in the handling and usage of drugs and medicines. His conduct must be guided by ethical considerations as well as by statutory controls. The word "drug" has acquired a sinister overlay, and it is used in such statutes as "The Misuse of Drugs Act", but it should be understood that all effective medicinal substances have power for good or ill, whether or not they are defined as drugs or medicine.

The Commissioners of the European Economic Community have sought to ease the free movement of pharmacists between countries in the Community. Directives have been produced, but they have not yet come into force. There are some differences of structure and emphasis between the degree courses which are the academic basis of qualification in different EEC countries. In France, for example, pharmacists undertake the analysis of normal constituents of body fluids, and their education takes account of this in its analytical, biochemical and physiological aspects. In Great Britain such analyses are undertaken not by pharmacists but by clinical biochemists or chemical pathologists.

To assist the free movement of goods in the Community, EEC directives on medicinal products have been made by the Council of Ministers and are now embodied in statutory instruments made in Great Britain as Regulations under the Medicines Act 1968. One effect of this is that each licensee manufacturing medicinal products must employ and designate a "Qualified Person" to be responsible for certifying that all the requirements of a licence have been carried out before a product is released. These requirements include analytical control, compliance with specification recording, labelling, storage and the coding of batches.

The Directives and Regulations require that a Qualified Person must have had at least two years' experience in qualitative analysis of medicinal products, quantitative analysis of active substances, and the testing and checking which is necessary to ensure the quality of medicinal products. He must before this have gained a degree in pharmacy, medicine, chemistry, pharmaceutical chemistry or biology. He must also produce evidence of adequate knowledge of a range of basic subjects, most of which are to some extent incorporated in the usual degree courses in pharmacy.

In many countries there are now a large number of pharmacy students in universities and polytechnics and this has led to the fear that there may be an over-production of pharmacists for the number of professional posts available. But the degree course in pharmacy provides a chemical, biological and experimentalist background which enables pharmacy graduates to find other forms of employment, and many have already found this possible and highly satisfying.

PHARMACOLOGY

Profile No. 10

PHARMACOLOGY

by

Professor George Brownlee
(Emeritus Professor of Pharmacology in the University of London)

Whereas Pharmacy is concerned with the preparation and supply of drugs, Pharmacology is predominantly concerned with drug action. The precise scope of Pharmacology, which comes from the Greek word "Pharmakon" for a curative drug, is debatable. Thus in 1937 A.J. Clark, then Professor of Pharmacology in the University of Edinburgh, defined it as "the study of the manner in which the functions of living organisms could be modified by chemical substances". This is certainly a broader definition than that used by the doctor, dentist or veterinary surgeon. It may be noted that Clark's definition includes all living matter whether plant or animal, whether single cells or multicellular structures. This implies that there is a common basis for the life processes to be found in many seemingly different kinds of cell.

Modern pharmacology had its origins in Germany in university departments attached to medical schools and was stimulated by the rapid growth of physiology in the early part of the nineteenth century. By the middle of the century they were using the experimental methods needed to determine the effects of the drugs and poisons which were known to them. At the same time several organic chemists had shown that it was possible to make in the laboratory substances which it had previously been thought could be made only by living organisms. The new science of biochemistry that emerged concerned itself with the chemical nature of life. Thus three branches of science, pharmacology, physiology and biochemistry, aided each other's development. In particular the success of the pharmacologists in elucidating the mode of action of drugs on living tissues enabled the biochemists to use the drugs as tools to allow them to probe deeper into the metabolism of living cells. This self-fertilising process has continued and has been a major feature of the continuing collaboration between pharmacologists and biochemists.

One of the first university departments of pharmacology was established in Edinburgh in 1768. A major interest of research in pharmacology for many years thereafter concerned the interaction of drugs and in particular the way in which one drug counteracts the actions of another. A vivid description of experimental pharmacology was given in 1855 by Robert Christison. He was interested in the effects of the Ordeal-Bean of Old Calabar and after describing the effects on rabbits he tells how he chewed and swallowed a quarter of the surprisingly tasteless bean. Some twenty minutes later he thought it time to bring the experiment to an end. "Being now quite satisfied that I had got hold of a very energetic poison, I took immediate means of getting quit of it, by swallowing the shaving water I had just been using, by which the stomach was effectively emptied". The active ingredient of the bean was later shown to be an alkaloid physostigmine, the action of which could be counteracted, or antagonised, by another substance, atropine. The explanation of the phenomenon then uncovered was that physostigmine interacted with the enzyme that normally destroyed acetylcholine (the agonist) that fulfilled a crucial role in the physiology of the body and that the antagonist competed for the receptor and thus prevented the action of the agonist.

The period between the two world wars was remarkable for two fruitful developments in pharmacology. First, the biological standardization of drugs, and second, the discovery of chemotherapeutic drugs.

The discovery in 1921 of insulin by Banting and Best promoted the need for standardization for here was a substance extracted from beef or pig pancreas which had a life saving effect when administered to human diabetics. In order for a diabetic to lead a normal life the amount of insulin injected must be carefully controlled and hence it is essential to know the potency of the preparation and highly desirable that the same criteria of potency should be used world wide. The precise chemical structure of insulin was not known until 1955 so that the purity and potency of insulin preparations could not be determined by chemical means. Thus a biological assay had to be set up in which the activity of a sample of unknown potency could be compared with that of a stable reference standard. The bioassay of insulin involved the effect of administering a preparation to mice. The insulin caused a dose dependent fall in the concentration of the blood sugar of the mice and the effect of this could be observed. Slowly a whole range of bioassays were evolved for many different drugs and in the course of their development many interesting fundamental observations were made on the action of drugs.

A valuable dividend gained from these bioassays was the appreciation that reproducible physiological responses could be measured in a number of different tissues as a response to a single stable chemical substance. Thus with insulin the bioassay in mice could be duplicated by one involving the effect of insulin on the uptake of glucose by the isolated rat diaphragm incubated in a test-tube. In other instances the effect of a drug on the length of smooth muscle suspended in a bath could be measured. The inference from this discovery was that the means were available to check the activity of a whole range of chemicals as potential drugs.

The advent of the new chemotherapy in 1935 brought about what has been colourfully, but accurately, described as the therapeutic revolution. Paul Ehrlich had discovered the therapeutic effect of Salvarsan in 1905. This was effective against the spirochaete that causes the venereal disease, syphilis, and was also active against the trypanosome that causes sleeping sickness, but it was inactive against the bacterial infections of man. When, in 1935, Gerhard Domagk made his very great discovery of the control of streptococcal infections with prontosil rubrum, the significance of this demonstration of a true causal chemotherapeutic agent dawned on pharmacologists. The collaborative effort required of the chemist, pathologist, microbiologist, pharmacologist and others was faced and met by the emerging pharmaceutical industry. Within months of the demonstration of the effective use of prontosil its active constituent, sulphanilamide, was isolated, and the bacterial range of effectiveness quickly extended, first by sulphapyridine (1938) and later by other substances which increased further the therapeutic range and reduced the toxicity.

Penicillin, found by Fleming in 1928, was first made available in 1942; streptomycin (Waksman, 1943) was found to be effective in tuberculosis; chloramphenicol (1949) was the first broad-spectrum antibiotic. The protective tests were simple to do; progress was rapid; the products were life-saving. The revolution had begun.

By 1950 the pattern had been repeated with anti-histamine drugs, corticosteroids, diuretics and, with far reaching effects, the antidepressant drugs. In many of these developments empiricism and simple trial and error tests were sufficient to develop potent drugs of great medical benefit but the subsequent studies on the mode of action were often responsible for uncovering previously unsuspected biochemical and physiological mechanisms.

These developments have continued to the present day with notable advances in the treatment of cardiovascular, nervous, and mental diseases with drugs, of which 95 in every 100 were unknown in 1950. What was now remarkable about the experiments was the application of ingenious new biochemical methods for identifying and labelling the sites of action of drugs.

In contrast with the preceding two to three decades, the rate of development of new drugs has now slowed to such an extent as to cause alarm. The reasons are largely economic. It costs about £4 million to develop a successful drug in Britain and Europe and probably more in America. The diseases that now make major demands on hospital care are slow to respond to research effort; often there is no simple animal model. Examples are rheumatoid arthritis, some kinds of cancer, viral diseases, the diseases involving inflammation of the connective tissues and mental diseases like schizophrenia. Much of the increase in cost arises from the statutory requirement to test the actions and toxicity of drugs in a variety of animals before they can be administered to man.

These properties are tested, if possible together with a clinically equivalent drug, in a restricted trial in man. After this stage all the available information about the drug is given to clinical pharmacologists and other clinicians in specialised hospitals for clinical trials. In such a trial, groups of patients take existing drugs, or placebos (inactive substances), or the new drug. Subjective judgement is minimised by a design in which the doctor or the patient does not know which group is which. A "double-blind" trial of this kind does not, of course, provide the knowledge about therapeutics or toxicity, which may yet be uncovered only when the drug is prescribed to a larger population. At all of these stages further tests may be deemed essential.

Before looking at the future prospects for pharmacology we ought to ask who the pharmacologists are and where do they work? The British Pharmacological Society was founded in 1931, and in 1946 issued its journal now named *The British Journal of Pharmacology*; the need for a Clinical Pharmacological section was met in 1970 and the *British Journal of Clinical Pharmacology* published in 1974. As in other scientific disciplines, the exchange of ideas between nations has become frequent in internationally arranged meetings and there have been particularly stimulating joint meetings arranged in recent years between the British and the Scandinavian, German, Italian and French Societies.

In the U.K. there are about 1400 pharmacologists listed at work today. Most (60%) work in the academic sphere mainly in Universities and Colleges of Technology, where in addition to teaching medical, science, dental, veterinary or pharmacy students, there is an active participation in research problems, often directed to the mode of action of drugs, sometimes called pharmacodynamics; a lesser proportion is to be found in Research Institutes, and in Government Departments. Here, interests range from aspects of drug control, such as the potency of drugs prepared from natural sources like insulin or certain antibiotics, to fundamental research. Investigations arising from the toxicology of food additives and food contaminants are further examples.

Many pharmacologists are employed within the pharmaceutical industry. Their published contributions to basic research have made a substantial contribution to the advancement of the subject. Some are concerned with the biological standardization of drugs of natural origin, and with the toxicology of new drugs. In the same context, most of the animal work on the absorption, distribution, and excretion of drugs, called pharmacokinetics, is done within the industry or within commercial specialised institutes. The modern industry of food additives, colours, flavours, sweeteners, acids, solvents, emulsifiers, improvers, and preservatives draws its toxicological expertise either from its own resources or from specialised institutes.

There is a growing number of medically qualified pharmacologists involved in clinical pharmacology. Their research work is concerned with how drugs produce their beneficial or toxic effects in man and with the factors that influence these effects, like absorption, plasma concentration, protein binding, half-life, metabolism and excretion. They have the specialised knowledge to make the controlled clinical comparisons of the effect of administration of two or more drugs. Some have special interests in the chemical substances introduced into the human body as food additives or environmental contaminants.

In a profile of this kind comment on the exciting prospects for pharmacology must be brief.

A major area of current research is concerned with the nature of chemical transmitters in the brain. As in the past this enthusiasm derives its thrust from the methods used by biochemists and physiologists and from the use of drugs as tools to elucidate the nature of nerve transmission. Rapid advances may be expected in the explanation of disturbances in brain function which underlie the common disorders of epilepsy, senile dementia, schizophrenia, depression, drug dependence, alcoholism and obesity. Useful clues to the chemical mechanisms involved in learning, memory, mood and sleep have been found from empirical testing and from drugs used clinically.

Other areas in which it is hoped to make real progress is in the chemotherapy of viruses and cancer. With viruses there is a very real problem in devising drugs because of the parasitic nature of viruses. Since they rely so much on the metabolism of their host cell for their existence it is difficult to devise an antiviral drug which is not toxic to the host. In the chemotherapy of cancer the drugs at present being used, in some cases rather effectively, are disappointingly unselective in terms of their effect on the tumour cells and the normal cells of the body. This explains the many undesirable side effects that result from the use of cancer chemotherapy. Research proceeds to discover drugs which are more selective for the cancer cells.

TOXICOLOGY

Profile No. 11

TOXICOLOGY

by

Dr W. Norman Aldridge
(Deputy Director of the Medical Research Council Toxicology Unit)

Toxicity is the response of a living organism to a poison. Such a definition, without qualification, includes all living organisms and all poisons. Almost by accident and tradition some studies which fall into this general definition are excluded. Thus, although penicillin is toxic to certain bacteria, the phenomenon is not normally encompassed by toxicology. On the other hand there is no doubt that the undesirable side effects of penicillin which affect some people could be classified as toxicology. The most commonly accepted definition is that toxicology concerns those effects of chemicals which affect the health and well-being of man.

Irrespective of its origin, a poison is defined as any substance which injures health or destroys life when it is introduced into or absorbed by a living organism. The toxicity of a substance consists of its capacity to do so and the toxic hazard represents the probability that it will lead to injury or death.

The view that natural materials are beneficial or harmless, and synthetic chemicals harmful is a commonly held, if naive, misconception. A moment's thought will show that many poisons are of natural origin, many occurring in plants and fungi. Man has always been exposed to these natural poisons, but before the development of modern medicine there were many other hazards to worry about, such as infectious diseases. Now, the population of the world is much larger and a high proportion live in large groups in cities. These changes have resulted in many technological changes: the concentration of specialised industries in particular areas, the transport of chemicals and other materials over large distances, the use of biologically active substances such as pesticides to increase food production, etc. All of these changes have tended to increase the exposure of man to a variety of new chemicals. Some of these are rather persistent in the environment and their release in one part of the world leads to their appearance elsewhere. Since scientific and technological advances now allow us to detect these substances in small amounts in the general environment there is often much worry that exposure to small amounts of some chemicals for a long time will bring about an unexpected epidemic of diseases many years later. Although it seems probable that much of this worry is misguided, with our present knowledge it is often impossible to give a definitive opinion about the hazards from such exposure.

Exposure of man occurs during medical treatment, at work, or in the general environment. There is no difference in principle between the adverse effects resulting from the administration of drugs to treat a disease and other forms of exposure to chemicals. Thus, the practical need for the reduction of risk or harm to man concerns the development of knowledge about the interaction of chemicals with complex biological systems.

The conventional way to assess the toxicity of a chemical is to administer it to animals and then examine them for untoward effects. This provides information about the toxicity to a particular species of animal, often mice or rats. Decisions are then made about the use of the chemical by man after allowing a large safety factor. This is, of course, an empirical approach which, because the safety factor is large enough, allows decisions to be made which do not greatly increase the risks from exposure. However, accepted testing procedures have been found inadequate on many occasions. An example was the thalidomide tragedy when a sedative drug, given to pregnant women, resulted in the birth of many deformed children. Another more recent example is the use of practolol for the treatment of high blood pressure; the prolonged daily administration of this compound has led to serious side effects in a small proportion of patients. It is clear, therefore, that the routine testing procedures may not discover all potential toxicity to man. In many cases the safety factors imposed, because of many uncertainties, also result in useful drugs and chemicals being rejected. In addition the very high cost of conventional animal testing methods is tending to restrict the development of new compounds. New reliable, cheaper and more rapid tests are required.

Thus, the science of toxicology is in a state of flux but is now developing rapidly through research. Looked at from a biological point of view, the toxicity of a substance depends on many variables. It must be qualified by many factors such as the physical characteristics of the substance, the route of absorption (mouth, lungs, skin or injection), and whether the effects produced are reversible or irreversible. In addition to these qualifications toxicity is not a general response. In almost all cases toxic chemicals produce special kinds of toxicity, for example damage to a particular organ, symptoms resulting from the prevention of the activity of a particular enzyme, etc. In a general sense, the reason for this is that mammals consist of highly organised biological systems containing many control points. Many of these control systems are operated by a messenger (often a small chemical molecule often called a hormone, neurotransmitter, etc.) operating on a macromolecule (an enzyme, nucleic acid, receptor, membrane). Thus, in an entirely comparable way opportunities exist for the derangement of such control systems by their interaction with synthetic or natural chemicals. Problems in toxicology are, therefore, fundamental problems in biology. The development of understanding of the mechanisms of different kinds of toxicity caused by chemicals is necessary so that the routine testing procedures may be up-dated and made less empirical and more quantitatively applicable to man. Knowledge of all mechanisms of toxicity is required, but especially urgent are those involving cancer, mutations, immune processes and the nervous system.

Toxicology is a science which requires the integration of work by scientists trained in many disciplines. Those which quickly come to mind are biochemistry, chemistry, medicine, pharmacology, physiology, pathology and zoology, though expertise from other specialities is often required. This has important consequences for education in the subject, and particularly when considering the vast range of activities from routine experimentation to research aimed at establishing the primary molecular interaction leading to a particular disease. Among the technical posts there are many opportunities for those leaving school with a knowledge of several scientific subjects; further education in technical colleges is normally required. In the U.K. at the present time it is not possible to take a primary degree in toxicology so that one should first obtain a training in one of the basic disciplines, e.g. biochemistry. Depending on the aims of the student with a first degree, posts can be obtained in many institutions, industrial and governmental; sometimes further education is required; probably the best option is a postgraduate Master's degree of formal course work followed by a period in a suitable research laboratory.

At the present time there are opportunities for careers in toxicology in industry, governmental laboratories, regulatory bodies, universities and research councils. There is room for able students interested in various scientific disciplines. Toxicology is a subject in which the solution of practical problems often requires knowledge of the mechanisms involved in normal biological systems. This is an interesting and exciting intellectual challenge.

Further reading

Aldridge, W.N. (1973) New insights needed in toxicology. *New Scientist* **59**, 492-493.

Aldridge, W.N. (1980) Chapter 40, "Hazards from chemicals in industrialized societies" in *A Companion to Medical Studies*, Vol. 2, 2nd edition (ed. R. Passmore and J.S. Robson).

Baker, E.L., Zack, M., Miles, J.W., Alderman, L., Warren, McW., Dobbin, R.D., Miller, S., Teeters, W.R. (1978) Epidemic Malathion poisoning in Pakistan Malaria workers. *Lancet* **1**, 31-33.

Boyland, E. and Goulding, R. (eds.) (1974) *Modern Trends in Toxicology*, Vols. 1 and 2. London, Butterworth.

Hall, R.H. (1974) *Food for Nought: the decline of nutrition.* New York, Harper and Row.

Lawrence, D.R. (ed.) Drugs: development and use. *British Medical Bulletin* **26**, 185-259.

Sharratt, M. (1977) Evaluation of the safety of chemicals, in: *Current Approaches to Toxicology* (ed. B. Ballantyne). Bristol, Wright.

World Health Organisation (1971) Alternative insecticides for vector control. *Bulletin of World Health Organisation* **44**, 1-470.

NUTRITION

Profile No. 12

NUTRITION

by

Professor Arnold E. Bender
(Professor of Nutrition and Dietetics in the University of London)

Compared with the basic sciences such as chemistry and physics, the study of nutrition is relatively new – the first degree in the subject in Europe was established only 25 years ago at Queen Elizabeth College in the University of London. Even now Surrey is the only other University in Great Britain with a first degree in the subject. Nutrition, like medicine, is an integrated subject built on the basis of chemistry, physics, biochemistry, physiology and microbiology among the pure sciences, and also upon sociology and psychology among the behavioural sciences.

Nutrition can be defined in two parts; the study of food in relation to man, based on the pure sciences, and the study of man in relation to his food, based on the social sciences. Both aspects are equally important and are illustrated in the diagram.

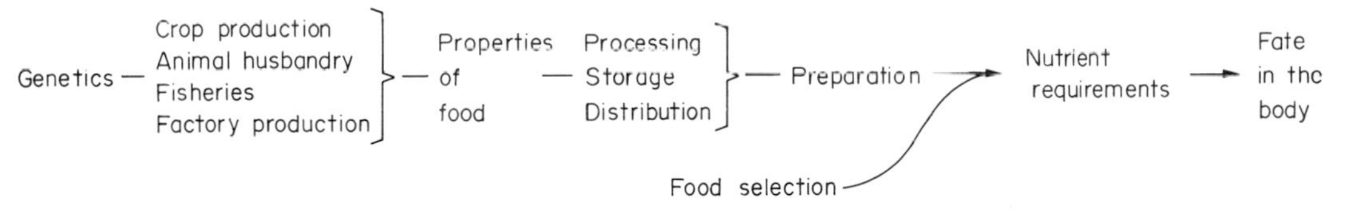

Field of nutrition

We start with food production, not only with the farmer and the fisherman who produce food, but with the geneticist who breeds improved varieties. A partial solution to the problem of any protein shortage, for example, is to breed varieties of plant which are rich in protein of high quality. This objective has been achieved for many of the cereals and is being developed in the area of legumes. Food production is no longer limited to the soil and the sea and it is possible to product yeast, bacteria, moulds and algae on a factory scale, both for human food and animal feeding – the animals, in turn, providing us with food.

After the food has been produced the study of its properties, chemical and physical, biochemical and biophysical, is the realm of the food scientist. We need to know the properties of foods in order to arrange the best conditions for storage and treatment. Plant foods in particular are metabolically active even after they have been harvested; changes take place in their composition, particularly with respect to the proportion of starch and sugars.

This results in changes in texture, flavour and physico-chemical properties which alter the way in which these raw materials are converted into attractive foods.

The next stage is the processing of the raw materials into the wide variety of foods that are now commonly available. This is necessary for several reasons; because the raw materials produced by the farmer are often not very palatable, to increase the variety of foods in our diet, and, most important, to preserve them. Although man has salted, smoked and dried his foods for centuries, before the days of modern technology there was always a shortage between harvests. Now that we are able to bottle, can, cool, freeze and dry foods on a vast scale and have fast international transport we may have food in abundance at all seasons – so long as the consumer has the money to buy it.

The nutritionist is not a farmer, a food scientist or a manufacturer but he plays a part in all these areas; unless each stage is investigated from the nutritional point of view and is adequately controlled the state of human health may be affected.

After all this the food has to be prepared for consumption, whether at home or in institutions such as schools, hospitals, factory canteens or hotels. Unless care is taken, part of the nutrients may be destroyed and unless the food is prepared in an attractive manner it may not be eaten at all. The study of the selection of the actual food that is eaten is the province of social nutrition. Why do people eat what they do, and if their choice is nutritionally poor how can it be altered? Many factors influence our choice of foods, tradition, culture, habit, religion and price among them. When an immigrant arrives to settle in a new country he rapidly learns to alter his clothes, so as not to appear too different from his neighbours, but he takes much longer to change his food habits. Some remnants of the original cultural background persist for centuries. Many Americans long since settled in other countries continue to eat turkey, cranberry sauce and pumpkin pie on Thanksgiving Day; the Englishman in Australia will take his Christmas pudding onto the hot, sunny beach; the central European will prefer pickled olives, chopped liver, pickled herring and salami. The consumption of meat, and indeed the type of meat, will be dictated in many instances by religion, as will the consumption of alcohol and various combinations of dishes, and fasting and feasting. Obviously, the strength of such dietary restrictions will depend on the degree of religious conviction.

Then the next stage is to ascertain how much of each nutrient people require – how much energy will satisfy our needs without causing us to get fat or lose weight, how much protein, and of what kind, how much fat and again of what kind, how much of each of the 13 vitamins and 20 mineral salts. Finally, we come to the fate of these substances in the body – a study of their biochemistry impinging on medicine.

So nutrition is a very wide field depending on a host of other sciences and disciplines. One example can be taken to illustrate the ramifications and interrelationships of this study of food. In some countries chicken is held in great esteem, especially in Malaysia, but even in Great Britain, because chicken was once a relatively expensive food it is still regarded by the older generation as a symbol of prestige. In an attempt to reduce the cost of production, agricultural economists pointed out that a chicken, like other animals including man, grows rapidly only during the early stages of life. Chickens grow fast for the first ten weeks and after that their growth slows so much that it is not really worth giving them all that food for so little return in extra meat. It was therefore suggested that if chickens were killed at ten weeks of age instead of twenty we would not only have two chickens where one grew before but they would cost less in terms of food provided to them – the broiler chicken was "invented".

Then the chicken farmer asked for nutritional advice on the best way to prepare the perfect diet for the most rapid growth for least food. Later the geneticist pointed out that if feed conversion efficiency was to be the major factor we were growing the wrong types of chickens and so a suitable strain was bred. Finally, the nutritionist determined whether the nutritional value of these younger chickens, which do not have as much flavour as older ones, are as good nutritionally – they are.

So the chicken of today plays a major part in supplying some of our nutrients, is cheaper, and serves as an example of the partnership between many disciplines.

Practice of nutrition

The work of the nutritionist can be grouped under four headings: bench, blackboard, desk and location.

1) **Bench.** Many nutritionists carry out research to determine the relation between diet and disease, or the interrelationship between the nutrients in maintaining good health. Some are engaged in more applied laboratory work in the food industry in providing new foods, dietetic preparations, baby foods or slimming aids.

There are many national research institutes involved in furthering our knowledge of major foods like cereals, milk, meat, fruits and vegetables to improve production, keeping qualities, acceptability to the consumer and so on, and there are always some nutritional aspects involved.

2) **Blackboard** clearly covers teaching – University, Polytechnic and Teachers Training Colleges, where the next generation of domestic science, nutrition and food science teachers, and caterers and dietitians are being trained.

3) **Desk** refers to advisory work in Government Departments such as Ministries of Food, Agriculture and Health, International Agencies and advising food manufacturers on the nutritional value of their food products, and in advising the public directly on health matters related to diet.

4) The **location** can be the concrete urban community where nutritionists are engaged in finding out what people eat in industrialised societies, how much and why, whether there are any nutritional problems in the infants, children, immigrants, the elderly or any other section of the community, and if so, how they can be overcome. The term also includes surveys overseas in developing countries where there is a great deal of malnutrition. This calls for the application of a whole range of measures, often with advice of the International Agencies such as the Food and Agriculture Organisation, the World Health Organisation, and the United Nations Children's Fund, which employ many nutritionists in what are generally termed Applied Nutrition Programmes.

Future developments

There is a growing awareness of the importance of the study of nutrition among consumers, governments and food manufacturers. More and more people are becoming interested in food and its relation to their health; they want to know more about the new types of foods being made available and the safety of artificial food colouring, flavouring and preservatives; they are asking what is in the food, and what effect it might have on them.

The growing interest of governments is reflected in legislation which increasingly requires packed food to carry nutritional information. Governments of developing countries are becoming more aware of their nutritional problems. The foods manufacturer obviously has to obey any legislation but also feels a growing responsibility in the whole area of food and health.

Finally, we are finding that more and more diseases are related to diet – diabetes, heart disease, cancer and diseases of the bowl. Some diseases can be cured or at least treated by dietary control.

Nutrition must be considered a growth area.

FOOD SCIENCE

Profile No. 13

FOOD SCIENCE

by

Professor Alan G. Ward
(Emeritus Professor of Food Science in the University of Leeds)

What food science is

In all human societies, raw foods undergo changes during the period between the gathering (of plants) or slaughter (of animals) and their consumption. Food science is concerned with these changes, whether they arise during the simple storage and cooking procedures of primitive people or as a result of the sophisticated processes of the modern food industry. All foods, unless adequately preserved, are unstable chemically as well as liable to attack by micro-organisms, insects and other animals. Many foods such as cereal grains need to be modified before consumption so that they become pleasant to eat as well as more readily digestible. These two ways of modifying raw foods, preservation and transformation by processing, are often achieved simultaneously, as is evident in many customary cooking operations.

The means used to preserve foods and transform them into more acceptable forms include (i) physical, such as subdivision, mixing, heating, chilling, freezing, drying and irradiation; (ii) chemical, including treatment with acids, preservatives, antioxidants, emulsifiers, etc. or with osmotic agents such as sugar and salt; (iii) biochemical, such as enzymes which are used in controlled reactions to break down or modify compounds, e.g. starch to sugars; (iv) biological, in which micro-organisms are usefully employed to effect desired changes, e.g. the alcoholic fermentation of cereals and fruit to give beer and wine, respectively, and the fermentation of milk by lactic acid generating bacteria as part of cheese and yoghurt production. Many different agents are used in the processing of food, and their considable variety illustrates the way in which food science has originated from a wide range of the basic sciences and still employs their knowledge and techniques. It shares this characteristic with medical science (in which we include human nutrition) and with agricultural science. Also in common with these two applied sciences, food science has a direct impact on human existence and health, and on family and social life.

The scientific study of agriculture is separated from that of food science by a boundary roughly established by the harvesting of crops, slaughter of animals or preparation of milk and eggs in bulk for despatch. This boundary is sometimes loosely referred to as being located "at the farm gate". The corresponding dividing point for fish as food is when the fish are caught, which is usually close to the time when they die. At the opposite end of the span of this science is the boundary between food science and human nutrition, which can conveniently be put at the point where the food is actually swallowed and starts to have its effect on the metabolism of the body.

The subject matter of food science, set within these two boundary points, includes the study of some processes which were originally carried out in homesteads and farms, such as cream separation, butter and cheese making, the curing and smoking of bacon and ham, and even brewing and wine making. The chemical analysis of foods for the content of the various nutrients is also part of food science. In industrialised food production, food science extends beyond manufacture proper to the scientific study of food distribution, including the effects of transport and of handling and storage in warehouses, supermarkets, shops and the home. It also has as its concern, although so far in a rather limited way, the techniques and effects on the food of meal preparation by the caterer or the cook in the home. Indeed many of the processes employed by the food industry began as scaled up kitchen operations, e.g. milling, bread making, jam making.

The last subject of study before food science gives place to nutrition is the human sensory reaction to food, the response to colour, to taste and to odour (which together give flavour), to texture and even to the sounds generated in eating. Progress here depends heavily upon assistance provided by the studies of sensory physiology and psychology.

Much of agricultural science and human nutrition, although differentiated from food science, is relevant to the work of the practising food scientist. Information needed from these sources concerning particular foods, or special problems being studied, can usually be successfully sought either in the literature or from the appropriate experts. In practice, major investigations in food science are normally carried out by teams of scientists, containing experts in the relevant aspects of agriculture and nutrition, specialists in particular basic sciences, and engineers concerned with practical applications. Food scientists may act as further specialists or may integrate the diverse aspects of the whole problem.

Food science and the basic sciences

(a) **Biology.** Raw foods, which are the materials used in food processing, originate from a rather restricted selection of the plant and animal species encountered on the earth. They necessarily bear the marks of their biological origins in their composition and structure and in the changes that can occur in them. So fruit removed from the tree still utilises oxygen and sugars to obtain energy in useable form. It continues to synthesize proteins and other compounds and so modifies its chemical contents. The changes may be slow, as in apples, which can be safely stored at temperatures just above 0°C for long periods, or faster, as in bananas, which ripen quickly and which cannot be chilled to too low a temperature without upsetting the delicate balance of reactions in the tissues. For meat, the cutting off of the oxygen supply to the muscle tissues at the animal's death causes the chemical changes to diverge more and more from those in the tissues of the living animal. The resulting changes in the meat after the animal's death have major consequences for its appearance, texture, stability and eating qualities.

These few examples of changes arising from the constitution of the raw food could be multiplied many times. They show that a basic knowledge of biology, with especial emphasis on the structure of the cell and its functioning and on biochemistry, is needed to understand the starting materials of food processing. Knowledge of the general principles should be extended by study of the specific tissues and structures which are used as foods — such as fruits, including cereal and legume seeds and nuts, the roots, tubers, leaves and stems of plants, the muscle tissue of mammals, fish and birds, avian eggs and milk, in particular cow's milk.

(b) **Microbiology.** Biology is also needed as a basis for the study of microbiology, which forms a major sector of food science studies. There is a need to understand the common features of the major divisions of micro-organisms, and for mastery of the most frequently used experimental techniques. But micro-organisms occur in or on food, and interact directly with it. This specialised area of food microbiology has several aspects:

(i) Spoilage organisms may destroy the value of the food. This happens when bread becomes mouldy or when fish is stored above the freezing point and is attacked by bacteria, giving it odours and tastes which make it unacceptable. Much potential food is wasted the world over through microbiological spoilage which could be avoided by the application of food science to its preservation.

(ii) Bacterial contamination and growth may make a food dangerous to eat, whether or not it has simultaneously caused the food to deteriorate in eating quality. This situation can arise because the organisms contaminating the food produce disease even if consumed only in small numbers (e.g. such pathogenic organisms as *Salmonella typhi*, causing typhoid fever). Illness can also result if harmful food poisoning organisms have grown freely in food which is kept in unsuitable conditions before eating. Consumption of large numbers of these food poisoning organisms upsets bodily functions, causing sickness, diarrhoea, dehydration and even death (e.g. gastrointestinal disturbance from various *Salmonella* species, or from *Clostridium perfringens (welchii)*). A further group of micro-organisms produces poisons (toxins) in the food before it is eaten. These toxins then cause illness or death (e.g. the powerful toxin from the growth of *Clostridium botulinum* which can occur in poorly sterilised canned meat, fish or vegetables).

(iii) Some micro-organisms can be harnessed to effect major desirable changes in foods. The changes serve the growth needs of the organisms and simultaneously change the food into a form appropriate for human consumption. Baker's yeasts (*Saccharomyces cerevisiae*) ferment sugars to ethanol (alcohol) and carbon dioxide in breadmaking. The carbon dioxide generated aerates the dough so that the bread acquires an open texture. Other yeasts, mostly different strains of *Saccharomyces cerevisiae*, are the basis of the production from various sugars of all the alcoholic drinks of the world, including beer, wine, cider and, after distillation, whisky, brandy, rum and so on.

So microbiological studies in food science are based on the principles of general microbiology but extend to the particular organisms which are important in food spoilage, food poisoning and food modification. A prime concern is how these organisms interact with food during storage and in the various food production operations. Resistance of the organisms to heat, their sensitivity to cold, their ability to grow if water availability (water activity) is restricted by drying or by addition of salt or sugar, and the effect on them of acidity and of preservatives are all relevant and important. Where micro-organisms are used as aids or agents in food processing, a further whole field of study is created, including the reactions they make possible and the optimal conditions needed for them.

(c) **Chemistry and biochemistry.** The chemical constitution of foods, which includes their content of nutrients, and the chemical reactions which occur during storage and processing are the province of food chemistry and biochemistry. As would be expected, raw materials derived from plants and animals are complex chemically. The presence of biological structures affects the ease with which the chemical compounds present can react with each other. Thus enzymes which, in the living organism, are carefully segregated from substances whose breakdown they catalyse, may become mixed with them during food operations. This may have adverse effects, as when a cut or peeled apple blackens, or it may produce useful changes, as when reactions resulting in blackening and flavour development are induced by the manipulation of green tea leaves in the manufacture of Indian tea.

The detection and estimation in foods of contaminants and of toxic constituents, both inorganic and organic, is carried out by highly developed analytical means. Similarly detecting the minute traces of very active flavour compounds sets problems for the analyst. Despite advances of the last twenty years, it is still true that the human nose – and also some animal noses – can detect traces of substances which are present in foods in quantities too small to be detected by the most sensitive chemical analyses.

A food scientist needs a thorough knowledge of the many classes of chemical compounds present in foods – proteins, amino acids, peptides, fats and oils, sugars and complex polysaccharides, organic acids, vitamins, the compounds providing colour, odour and taste, and many more. The materials used for containers and packaging, which are in contact with the food, are also important. But the relevant chemistry deals with more than just the structures of compounds and their reactions. The properties of solutions of small and large molecules, osmotic pressure, water activity and swelling are key topics. Surface chemistry, i.e. what takes place at the boundary of liquids, as at the surface of fat globules in milk, ice cream and mayonnaise or at cell boundaries and at solid surfaces such as those that enclose flour particles or sugar crystals, is another major chemical field which is very much involved in food science.

In making use of chemical knowledge, the problem for the food scientist is to select the parts that are relevant to the particular food subject being studied. This requires sufficient knowledge of how chemistry is organised, in order to find the information required.

(d) **Physics.** The extent to which physical agencies are used in the main methods of preservation and transformation of foods has already been emphasised. The high cost and short supply of sources of energy have stressed the need for efficient transfer of heat to food from the steam supply, or from hot water or electricity during processing. The heat used in this way can often subsequently be partially recovered and reused. Conduction, free and forced convection and radiation come into their own as subjects of study in food science.

Transfer of heat is also involved in the concentration of liquid foods – fruit juices, milk, sugar solutions – as well as in food drying. The need to avoid damaging or contaminating the food by overheating, corroded plant or in other ways is a constraint on the processes used. The latent heat for evaporation must be supplied economically and this need has prompted ingenious methods of energy saving, with no loss of quality in the product.

The interaction between radiation and food extends beyond the direct use of radiant heat and includes microwave heating by the absorption of electromagnetic waves not only at the surface but throughout the body of the food. Various forms of high energy radiation from radioactive waste or from particle accelerators have been used for killing micro-organisms and insect pests as well as for suppressing sprout growth in potatoes.

The flow of liquid foods and the mechanical behaviour of solid foods, which includes the way they break up in cutting, grinding and milling, have to be studied in order to understand many processes in food manufacture. The mechanical response of foods to chewing, in the very complex conditions which occur in the mouth, determines certain aspects of food quality.

Much of the physics of food and food processing is closely linked with engineering applications in food manufacture. The physical approach seeks accurate measurements and soundly based explanations, whereas food engineering is concerned mainly with the effective design and operation of processes on the factory scale.

(e) **Mathematics and statistics.** Qualitative changes play their part in food science but numerical and quantitative aspects are of key importance. The quantities concerned range from yields, efficiencies, contents of chemical compounds (including nutrients), to temperatures and heat flow in non-steady state heat transfer and the stress-strain components in mechanical behaviour. Variable raw materials and complex processes require the use of statistical methods both in the laboratory and in the factory. Sensory tests for taste, odour and texture, which are carried out by panels of testers, present special problems of interpretation requiring statistical treatment. So effective food science requires mathematical understanding, facility in the use of formulae and ability to use statistical methods as well as the services of statisticians.

(f) **Food science as an integrated approach.** This brief account shows that a formidable array of basic science must be mastered if food science is to be fully understood. Food science now has its own range of textbooks, monographs and research journals and contributes its own experimental results and theoretical explanations to the pool of scientific knowledge. Food science often requires the deployment of a range of techniques, drawn from different areas of science, to solve particular problems. We may consider, as an example, what is involved if we wish to test the effectiveness and safety of canning processes. We must consider the transfer of heat to the can by convection, and of heat within the can by conduction or convection. We must know what temperature is needed and for how long, in order to destroy bacteria and their spores in the can. Time and temperature also control the destruction of enzymes and other chemical reactions. We must know whether storage over a long period will affect the contents of the can, and whether it will corrode. The final factor is the human response when the can is opened and its contents are eaten. The characteristics of the final product depend on the changes in it which take place at particular temperatures. This example of the many pronged character of food science investigations and hence of the body of knowledge accumulated in food science could be multiplied over and over again.

Education in food science

Several educational routes are open to those wishing to become food scientists or to apply food science as food technologists. The most straightforward is to take a degree course in food science (or in the closely related but more practical food technology). The various courses available will have greater or lesser amounts of the basic sciences, according to the views of the particular university or polytechnic department.

The majority of food science students enter such courses having studied at school mathematics, physics, chemistry and biology.

A second approach to food science is to take a two-subject degree course, one subject being food science and the second either chemistry, physics, physiology, nutrition, biochemistry or microbiology.

Another route is to study one of the basic sciences to B.Sc. degree level and then to add to this a one-year postgraduate course (usually for an M.Sc. degree) in food science. Although it is more difficult to secure the proper balance between the elements which should make up a food science course in this way, there are compensations in the opportunity to delay the commitment to enter an applied science and in the depth of knowledge acquired in the first degree subject. A related route is to add research in food science (for M. Phil. or Ph. D. degrees) to a basic science first degree.

Further reading

Food Science, A Special Study in the Chemistry section of Nuffield Advanced Science. Penguin Books.

The Science of Food. P.M. Gaman and K.B. Sherrington. Pergamon Inter-

national Library. Pergamon Press.
Food Science and Technology. Magnus Pyke. John Murray.
Unilever Educational Booklets. Advanced Series. No. 3 – *The Chemistry of Proteins*; No. 4 – *The Chemistry of Glycerides*; No. 9 – *Micro-nutrients*; No. 10 – *Carbohydrates*; No. 11 – *Plant Protein Foods*. The above booklets from: Unilever Educational Section, P.O. Box 68, Unilever House, London EC4 4BQ.

ENDOCRINOLOGY

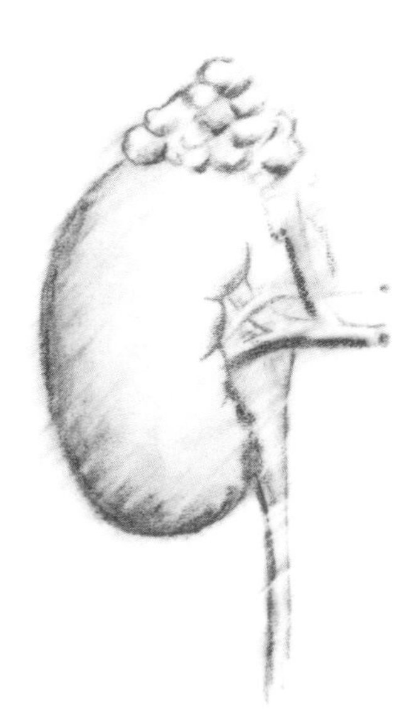

Profile No. 14

ENDOCRINOLOGY

by

Dr Bryan Hudson
(On the staff of the Howard Florey Institute in the University of Melbourne)

Introduction

Perhaps Nature is shown at her logical best in the functioning of the group of glands which comprise the endocrine system. The term "endocrine" is applied to those glands which make chemical messengers that are secreted into the blood stream and act on other tissues that are normally situated at some distance from the gland that secretes the chemical. These chemicals are called hormones, a term coined nearly 75 years ago by the distinguished British physiologist Starling from the Greek word meaning "I excite or arouse". By and large the mechanisms by which the secretions of these glands are regulated are nicely ordered and easily understood.

Let us put some names to these glands. First, there is the anterior pituitary gland which lies at the base of the brain. The location of this gland, which has been called "the conductor of the endocrine orchestra" is not accidental. We now know that although the anterior part of the pituitary gland is not itself part of the nervous system, it is to a considerable extent controlled by the brain. This was an exciting, relatively recent discovery. I shall refer again to the relationships between the brain and the endocrine glands.

Other endocrine glands are the thyroid, the parathyroids, the pancreas, adrenals and sex glands — the ovaries and testes. The thyroid is located in the neck and has as its principal secretion thyroxine, which is chemically unique among the hormones because it contains iodine. Thyroxine and its closely related hormones exert a general control over the metabolic activity of most of the tissues of the body, and are also important in their normal growth and development. Although the parathyroids are embedded in the thyroid gland, their functions are quite unrelated. The parathyroids secrete a protein hormone which regulates the metabolism of calcium, an essential element not only for the structure of the skeleton but also for the control of many metabolic activities at cell membranes.

While the main bulk of the pancreas consists of tissues which manufacture digestive enzymes that are secreted into the intestines and are essential for the digestion of food, there are parts of the pancreas that have an endocrine function. These small islands of tissue secrete the hormone insulin, necessary for the control of the metabolism of glucose which is an essential source of energy for all the cells of the body. In the absence of insulin glucose cannot serve as a fuel so most cells are deprived of energy and waste away. A deficiency of insulin leads to diabetes. Although the secretion of insulin from these pancreatic islets was discovered nearly 60 years ago, only relatively recently was it appreciated that they also secrete at least one other hormone, glucagon, which also plays a role in glucose metabolism.

The name adrenal is given to the two small glands that lie on the top of each kidney. The outer part of each of these glands is called the adrenal cortex, and life cannot be sustained without it for more than a few weeks. The cells of the adrenal cortex secrete steroid hormones, of which the predominant one is cortisol. Not all the actions of cortisol are properly understood, but it is known to effect the synthesis of protein in many cells of the body. Cortisol is closely related to cortisone which is widely known as a drug for the treatment of a number of inflammatory disorders such as rheumatoid arthritis and related diseases. We are still largely ignorant of how these adrenal steroids act in this way. Another important adrenal steroid hormone is aldosterone which regulates the metabolism of water and salts; if aldosterone is absent, salt (particularly sodium chloride) cannot be conserved by the body.

The inner part of the adrenal gland is called the medulla and it is surrounded by the cortex. The medullary tissues are quite different from the cortical cells in origin, appearance and function. The secretions of the medulla are adrenalin and noradrenalin. Adrenalin has been a household word for many years. Terms like "I am going to get my adrenalin going", or "He uses up too much adrenalin", give a clue to the nature of this hormone. It stimulates many body functions, increases heart rate, makes the hair stand up, supplies energy, and has been termed the hormone of "fright, fight or flight". This is how it acts when secreted by the adrenal into the blood stream. Adrenalin belongs to a class of biochemicals called catecholamines, which are found in many other parts of the body; for instance noradrenalin is present in the central nervous system where it acts to transmit nervous impulses.

Finally there are the reproductive glands, the ovaries and the testes, which have two functions. One is the secretion of sex steroid hormones; the other is the production of gametes ... ova and sperm. There are two classes of sex steroid hormones: those secreted by the testes, of which testosterone is the most important, and those produced by the female (ovary) of which oestradiol is the most potent. These hormones are responsible for the development of a number of sexual characteristics ... for instance, breast development and feminine contours in girls, and beards and broken voices in boys, as they each pass through puberty. Sex hormones also play a significant role in the development of gametes, and in the female for the preparation of the reproductive tract to accommodate the egg should this become fertilized.

More recent discoveries

So far the classical tissues of the endocrine system have been described. These have been known for at least sixty years. In the last twenty years or so, remarkable discoveries have been made about other tissues which have endocrine functions. Scattered along the upper part of the gastrointestinal tract, particularly in the stomach and the duodenum, are groups of cells which also secrete chemical messengers that act on distant parts of the gut. They cause them to secrete, to contract or to relax, and thus complement the nervous system in digesting and propelling foodstuffs along the gut. These are the gastrointestinal hormones. Indeed, the first chemical secretion recognized as a hormone by Starling in 1905 was secretin; others have been discovered more recently. These include gastrin, motilin, bombesin and cholecystokinin. Not all have been chemically characterized nor are their actions known or understood. This is still an area of active and continuing research.

The brain as an endocrine tissue

An exciting recent discovery in endocrinology is the relationship between the brain and the endocrine system. Attention has already been drawn to the anatomical association between the brain and the pituitary gland. The region of the brain beneath which the pituitary gland lies is the hypothalamus, which is associated with the pituitary in two ways: first, there is a connection by a series of blood vessels that carry blood from the hypothalamus to the anterior lobe of the pituitary. The second connection is by nerve fibres which originate from collections of neurones (nuclei) in the hypothalamus and terminate in the posterior lobe of the pituitary. It is of interest that the hypothalamus has nuclei which control thirst, appetite and body temperature, but there is no direct evidence that these are directly connected with the endocrine system.

Thus, the pituitary gland has two lobes which are quite different in origin, structure and function: the front or anterior lobe is glandular while the posterior lobe is composed mainly of nervous tissue. The anterior lobe secretes at least six hormones, four of which regulate other endocrine glands. These are called "trophic" hormones. This term is applied because they nourish or stimulate these glands. Thus the thyrotrophic hormone nourishes the thyroid gland and the adrenocorticotrophic hormone the adrenal cortex. If the anterior part of the pituitary is removed and the trophic hormones are absent, the nourishment of the thyroid, adrenal cortex and reproductive glands are cut off and these glands wither and atrophy. Note there are no known trophic hormones for the pancreatic islands, the parathyroids or the adrenal medulla.

Two other hormones are secreted by the anterior pituitary. These are growth hormone and prolactin. As its name suggests, growth hormone is related to growth; indeed, it is vital for growth in children and without it normal height cannot be achieved and dwarfism results (although not all dwarfs are growth hormone deficient). Prolactin is important as a hormone in women for breast development, when it acts in synergism with oestradiol – the female sex hormone – and after this to stimulate milk secretion in mammals following delivery of the new born. In lower species prolactin plays a role in salt and water regulation.

The posterior lobe of the pituitary is made up of nervous tissue, particularly the endings of nerve fibres that originate from nuclei within the hypothalamus. The bundle of nerve fibres that runs from the hypothalamus to the posterior pituitary forms the pituitary stalk. This connection gave a clue that the brain may make hormones. The important hormone in this instance is a peptide containing only a few amino acids linked together, the antidiuretic hormone (ADH), made by neurones in the hypothalamus and transported down the nerve fibres to the posterior lobe of the pituitary where it enters the blood stream. The hormone plays a crucial role in regulating the output of water by the kidney. We all share the common experience that when we are deprived of fluids we pass less urine, but that when we drink a lot of fluid copious amounts of urine are formed. Under conditions of fluid deprivation, hypothalamic neurones become more active, and the posterior lobe releases into the blood antidiuretic hormone which acts on the kidney to conserve water. When fluid intake is excessive the synthesis and secretion of ADH are switched off. These neurones are sensitive to changes in the composition of the blood that lead to a change of as little as 1% in the osmotic pressure. If there is a disruption of these nerve fibres, as may sometimes occur following head injuries or tumours in the region, ADH is no longer secreted into the blood, the control system breaks down and uncontrollably large volumes of urine are formed.

Hormones other than ADH are made in the hypothalamus and are carried by blood vessels which run from the base of the hypothalamus to the anterior pituitary (see Figure). These hormones either stimulate or inhibit

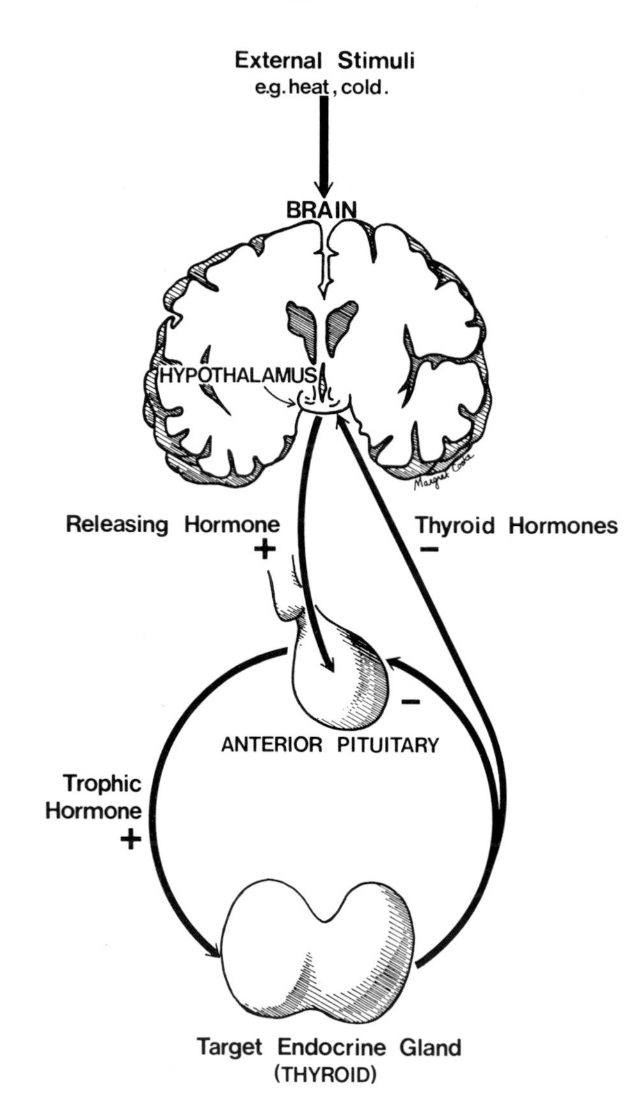

The figure is intended to portray the classical system for the control of many endocrine glands. In this instance the thyroid gland has been chosen as the example. External stimuli acting on the brain may be transmitted to the hypothalamus and cause the secretion of the releasing hormone which acts on the anterior pituitary causing it to secrete the trophic hormone. Note the autoregulatory aspect of the system: that is, the secretion of thyroid hormone "turns off" the secretion of releasing and trophic hormones by acting on the hypothalamus and anterior pituitary.

the release of anterior pituitary hormones. Those which were first discovered were the releasing hormones ... thyrotrophin releasing hormone and gonadotrophin releasing hormone. These are small peptides and are formed in the nerve cells. They are then secreted into the special blood vessels that run a short distance to the anterior pituitary where they act to release the appropriate trophic hormone. Not all substances formed in the hypothalamus are peptides. We have already noted that the adrenal medulla makes catecholamines, mainly adrenalin and noradrenalin. This latter amine is also made in the brain where it acts with a closely related amine, dopamine, as a transmitter of nervous impulses and thus perhaps stimulates the formation of hypothalamic peptides. Dopamine may also be secreted by hypothalamic neurones and regulate the secretion of some anterior pituitary hormones. The brain then, and the hypothalamus in particular, is very much a part of the endocrine system and plays a central role in coordinating the activities of many endocrine tissues.

Control mechanisms in the endocrine system

The comment has been made that the endocrine system is a nicely ordered one. The systems of endocrine glands are linked by a series of feedback loops which form servo mechanisms to regulate hormone levels within given limits. These may take different forms. There may be a direct control, as when ADH secretion is "turned on" in response to water deprivation and "turned off" when excess water is taken. Other examples are: when glucose levels in the blood reach a certain level the secretion of insulin is stimulated, and blood sugar falls; the effect of potassium on the glomerulosa cells of the adrenal cortex, or of calcium on the parathyroid. Another form of control exists between trophic hormones and their target glands. As shown in the schematic diagram the feedback loop may be made by hormones in blood acting on the pituitary itself or via the brain, probably both. There may be additional controls of the secretion of some hormones in the tissues themselves where there is a local or very short loop feedback system.

Current research: Mechanisms of hormone action

Until relatively recently we were able to give only rather crude descriptions of hormone action. For example, we knew that thyroxine caused increases in heart rate and body metabolism, and that testosterone stimulated hair growth and muscle development, but we knew little about the mechanisms at the molecular level.

A fundamental concept is that each hormone acts on its own target cells or tissues. Thus the trophic hormone for the adrenal cortex has no effect on the steroid secreting cells of the ovary or on the pancreatic islands. One reason for this exclusive action is the presence of specific receptors for a given hormone in a particular group of cells. These can be likened to locks which only take specific keys. Once the key (hormone) turns the lock (receptor) a door opens. In this analogy the opening of the door is the start of a train of biochemical events characteristic of the particular cell and hormone in question. For example, the male sex hormone acts on hair follicles in the face to stimulate hair (beard) growth.

Receptors for hormones may be contained in the membrane that surrounds the cell, or lie within the cytoplasm which means that these particular hormones must pass across the cell membrane to interact with their receptors. The ultimate action of most hormones is to stimulate the cell nucleus to make nuclear messages that will trigger the cell mechanisms to make a particular chemical (usually a protein) or a group of chemicals. The whole field of hormone action has been under intense study for the past 15 or 20 years, during which we have obtained some exciting basic new information about hormones.

Many references to the term genetic engineering can now be found in the daily newspapers and popular science magazines. Biomedical scientists who have been trained in endocrinology have made interesting and exciting contributions to this new science. For instance, in collaboration with biochemists and molecular biologists, endocrinologists have been able to synthesize the gene for rat insulin. This DNA has been transferred into the genetic apparatus of a common bacterium (*E. coli*) which is easily grown in culture. It is hoped before long to find ways of causing the bacteria to make rat insulin. There is no reason why they could not also make human insulin.

This new technology has important potential implications for the production of insulin and other hormones. At present insulin cannot be synthesized chemically on a commercial basis and the world's supply of insulin has to be extracted from the pancreas of pigs or cattle. In the foreseeable future it is possible to envisage that other hormones in short supply might be made by bacteria.

Relevance of endocrinology to human health and disease

Many advances have been made by basic research in endocrinology. These started nearly 60 years ago with the discovery of insulin and its use in the treatment of patients with diabetes. Before this discovery many of them died within weeks or months of the onset of their disease. Since then a whole series of hormones has been isolated, characterised and many of them synthesized and administered to patients with deficiencies of these hormones which in some instances would previously have proved fatal.

With increasing knowledge of the relationship between the brain, the pituitary and the ovaries, for instance, it has been possible to devise medications which either suppress or stimulate ovarian function. The oral contraceptive pill prevents development of ova. Other hormones stimulate ovulation in women who ovulate irregularly or not at all, and restore their fertility.

Knowledge of how hormones work has enabled us to understand better, and thus diagnose and give rational treatment to patients whose disorders are the result of abnormalities of hormone action. These successes would not have been dreamed of twenty years ago. Endocrinology is one field where basic and applied science can be fruitfully united and bring great benefits to mankind.

IMMUNOLOGY

Profile No. 15

IMMUNOLOGY

by

Professor Peter N. Campbell
(Professor of Biochemistry in the University of London)

We are all aware of the immune system in our bodies. Sometimes it seems to work to our advantage as when the first attack of an infectious disease gives a life time of immunity or when we are vaccinated against smallpox. Sometimes it seems to be a nuisance, as when we have hay fever and suffer the symptoms of an allergic reaction or when we need a kidney transplant. Memory, specificity and the recognition of non-self lie at the heart of immunology. We shall consider these in turn.

The first contact with an infectious organism clearly imprints some information, imparts some memory, so that the body is effectively prepared to repel any later invasion by that organism. This protection is provided by antibodies evoked as a response to the infectious agent which behaves as an antigen. Interaction with antibody leads to elimination of the antigen. The first response to antigen causes the production of a population of memory cells so that when the body is challenged again by the same infectious organism there is a more rapid and more abundant production of antibody so that no symptoms of an infectious illness result. Vaccination utilizes this principle of an adaptive immune response by employing a relatively harmless form of the antigen, e.g. a killed virus, as the primary stimulus to imprint "memory".

An example of specificity is the fact that the establishment of memory or immunity by one organism does not confer protection against another unrelated organism. After an attack of measles we are still susceptible to an attack of mumps. Thus the body can differentiate between the two organisms; the antibodies produced as a response to the two organisms are subtly different.

This ability to recognize one antigen and distinguish it from another goes even further. The individual recognizes what is foreign, i.e. what is "non-self". If the body could not discriminate between "self" and "non-self" it might form antibodies which could attack its own components with harmful results. The explanation is that even before birth circulating components of the body reach the cells which are responsible for the synthesis of antibodies, and these components are recognized as self. A permanent unresponsiveness or tolerance is created so that in the adult there is an inability to respond to self components.

The adaptive immune response of higher animals has evolved to provide a more effective defence in that appropriate immunological cells concentrate their energies on the particular agents infecting the body at any one time; the specific antibodies kill the organisms which greatly speeds up their disposal by the phagocytic cells of the non-specific immunity systems which are a feature of lower organisms.

Much of what I have described has been known in general terms for a long time. Thus immunization against smallpox by giving a mild infection to protect against a severe attack was introduced into England from Turkey about 1720. In 1798 Jenner proved that vaccination with cowpox was as effective and much safer. The years 1880 to 1910 were a period of great activity during which it was shown that the administration of many substances could cause the formation of antibodies. The antibodies were shown to cause the aggregation or clumping of the cells associated with the specific antigen but to do this required the presence of a substance which is present in fresh normal blood serum but is lost after storage or gentle heating. This substance was called "complement".

Between 1910 and 1935 progress was relatively slow and in the main was confined to the development of safer vaccines to protect people against infections such as diphtheria. Much was learnt about the properties of antibodies before anyone knew what they were, let alone what cells made them.

During the 1939-45 World War immunization against a wide range of diseases was used and great progress was made in the fractionation of serum proteins so that the fraction containing antibodies could be purified. Problems of skin grafting burned patients emphasized the already known, but unexplained, fact that, as a rule, only one's own skin can be grafted. The rejection of "foreign" skin was shown to be due to immunological mechanisms.

While antibiotics are immensely important in combating many bacterial diseases they do not cure all of them and do not cure diseases which are due to viruses. This emphasized the importance of studying the natural protective mechanisms and of improving immunization. Two major triumphs were the vaccines against whooping cough and the Salk vaccine against poliomyelitis.

These advances, although based on hypotheses about the nature of the immune response, were largely developed on a purely experimental (empirical) basis. A coherent theory was lacking. Between 1958 and 1960 this changed, firstly through the demonstration of tolerance, already referred to, and secondly through the emergence of the Clonal Selection hypothesis to explain the biology of immune responses. This stated that there exists in mature animals a population of cells, guessed to be lymphocytes, each precommitted to recognize and respond to one or two antigens. Interaction between the cell and the specific antigen would stimulate some of the responding cells to multiply and others to make antibody. The immune response and the phenomenon of memory could be regarded as reflecting a change in the number of precommitted cells.

Lymphocytes were then shown to be the cells involved as predicted in the theory. The total population of lymphocytes in a human adult is about 2,000 billion and weighs close to one kilo. The cells die and are replaced continuously, the turnover leading to the emergence of about a million new lymphocytes a second.

In mice and man it occasionally happens that the descendants of a single lymphocyte multiply out of control giving rise to leukaemias, in one form of which – myelomas – they secrete large amounts of exactly similar antibody molecules. Analysis of such myeloma proteins has allowed the differences between antibody molecules to be dissected. The elucidation of a general structure for antibodies was finalised in 1960.

Knowledge of the structure of complement, already referred to, and the way it interacts with antibodies has advanced rapidly. Complement provides a means whereby antibodies with innumerable specificites on combining with antigen can activate a much smaller number of biological processes. These vary from stimulating phagocytic cells to ingest and destroy the invading bacteria, to causing local inflammation and the pathological changes of diseases such as nephritis and rheumatoid arthritis. Another consequence of the interaction of antigen and a particular type of antibody is to cause the symptoms of hay fever and asthma. Further, in abnormal circumstances the rule that antibodies to self are not formed breaks down and so called "auto-antibodies" are formed which can react with body components and produce diseases like pernicious anaemia and rheumatoid arthritis.

All the lymphocytes in the body originate in the bone marrow cells. Initially these cells are in the precursor state and are all exactly alike. It was realised in the early 1970s that they then split (in roughly equal numbers) into two distinctive types: one called B-cells and the other T-cells. The precursor cells that become T-cells migrate from the bloodstream to the thymus gland, which is their exclusive finishing school. Those that become B-cells go to various lymphoid tissues (commonly known as glands which swell on infection).

The B-cells are the precursors of the antibody forming cells already discussed. The T-cells do not secrete antibodies but possess specific surface receptors by which they can recognize antigens and respond to them. T-cells can themselves differentiate into cells which directly kill other cells which have on their surface antigens which the lymphocytes can recognize, in particular virally infected cells. They also recognize "transplantation antigens" which distinguish one individual from another. Thus the characterization of T-cells has been important in helping us to understand the basis of the rejection of skin grafts and organ transplants. Progress in this respect has been made by the development of drugs which suppress the immune response, the so-called immunosuppressant drugs. The effect is, however, non-specific so that while the transplanted organ may not be rejected the patient is particularly vulnerable to infections.

An important new development in immunology is our ability to fuse myeloma cells, those cells which produce myeloma proteins in leukaemia, with other antibody producing cells from purposely immunized animals in such a way that a hybrid myeloma cell (inevitably termed "hybridoma") secretes the specific antibody programmed by the cell with which it was fused. By ingenious techniques suitable cells can be isolated and propagated in a test tube. It is thus in principle possible to obtain unlimited quantities of antibody completely specific for a single antigen. Once the technique has been adapted to human antibodies we may even expect specific antisera for use in man to be factory produced.

While most universities include the study of immunology in science degrees in biology few as yet offer a degree which specialises in immunology. This is probably wise since you will have realised that an understanding of immunology depends on a sound basis of cell biology, biochemistry, physiology and genetics. It is an exciting area of biology which clearly has great potential both for our understanding of biological phenomena and at the practical level for the preparation of better vaccines and for overcoming the problems of organ transplants.

Further reading

This profile is based in part on an article in *The Biologist*, February 1980, by Professor J.H. Humphrey, on "The Growth of Immunology" and on "Essential Immunology", 4th edition, by Professor I.M. Roitt, Blackwell Scientific Publications, Oxford.

GENETICS

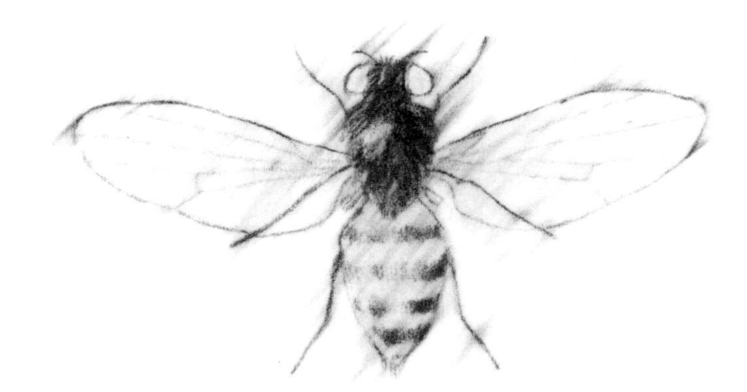

Profile No. 16

GENETICS

by

Professor Joshua Lederberg
(President of The Rockefeller University)

The science of living things is too complicated both in method and in objective to yield to tidy classifications. It is wrong to think of genetics as a distinct branch of biology: this would imply the existence of other compartmented branches. Rather it is a particular way of looking at almost every aspect of biology. Genetics is centered on the question of heredity: how does a given organism derive its characeristics from its ancestry and pass them on to its progeny. It thus focusses on the intrinsic quality of an organism. This in turn can be understood only in relation to how its development responds to the extrinsic environment that it experiences.

The methods used by geneticists embrace those of all the other biological specialities; and the object organisms that illustrate genetic principles range from viruses to the human. Particular emphasis is placed on morphological methods that display the chromosomes, and chemical ones for DNA, since these objects are the material carriers of information in heredity. Organisms used for genetic study may be chosen primarily for their convenience in the laboratory, e.g. bacteriophages (bacterial virus), *E. coli*, drosophila or inbred mice. Or, as in the case of plant breeding or human genetic disease, vitally important economic and medical practical objectives may be at stake.

Most geneticists thus find themselves also attached to one or even several other biological specialities covered elsewhere in this volume. On the one hand the geneticist will probably have specialized on a given group of organisms (see "Handbook of Genetics" by King in reading list). On the other hand he or she may have in mind any of the themes like molecular biology, physiology, nutrition, ecology, behavior, medical sciences or evolution. The methods of biochemistry, biophysics and structural biology can hardly be avoided. Some aspects of genetics also make heavy call on demographic and statistical analysis.

Genetics was founded on the observations made by Mendel over a hundred years ago. Using the garden pea for his observations, he found that hereditary traits followed regular rules of transmission: the now familiar Mendelian Laws. We now know that these laws are manifestations of the usual behavior of chromosomes during the "reduction divisions" of gamete-forming cells in higher plants and animals. Many exceptions to these rules are now known; but in 1865 they enabled Mendel to postulate the existence of the genes as relatively autonomous particles underlying heredity.

For many decades genes were invisible and beyond chemical analysis. Since the identification of DNA as the genetic material, by Avery, MacLeod and McCarty, in 1944, the task of genetics has been to integrate the study of the gene into the overall framework of the chemical biology of the cell. This integration, now largely successful, still presents formidable challenges to all of experimental biology.

Future challenges

Genetics has moved so rapidly in the past decade that it is unlikely that we can reliably foresee even the most fruitful and rewarding directions of advance over the next 20 years. The scientific objectives of genetics are scarcely distinguishable from those of molecular biology as applied to DNA, the structure of chromosomes, and the mechanisms by which genetic information is transcribed and translated into protein and the formation of more complex structures. Genetic aspects of evolution require an even more complex synthesis of these principles into the global panorama of evolutionary development. At the present time, evolutionary theory is still founded mainly on the most simplified models of genetic change. Thus it takes little account of more complex forms of variation. Some of these involve the transfer of blocks of genetic information from different species, and other modifications of DNA more complex than the change in a single nucleotide base. The limitations here are both in the inherent chemical complexity of DNA changes themselves, and the almost intractable mathematical complication of the comprehensive theory to model the changes. For these reasons the logical completeness of our present theory of evolution is still controversial. To put this another way, can we demonstrate that four billion years was enough time for evolution from the primeval ooze to contemporary humankind?

This question also opens up extensions of genetics to such issues as (1) what were the actual steps in the original evolution of life on earth from inorganic matter? and (2) what can be said about the distribution and diversification of life on other planets in the solar system and beyond? I have no doubt that many more missing links are still to be found among the species inhabiting earth; except for the most primitive initial exploration of Mars, of course, questions of the extraterrestrial distribution of life (exobiology) hardly go beyond intelligent and informed speculation.

As to the connections of genetics with developmental biology: rather detailed information is now available about the mechanisms that regulate gene expression in microorganisms. At least a number of provocative model systems have been put forward. There is nevertheless grave doubt whether these models account for gene regulation in eukayotic organisms. Hence the fundamental problem of embryology – how cells differentiate – remains obscure in molecular terms. The recent illumination of DNA-switch mechanisms in bacteria must make us wary even of insisting that the DNA in the nuclei of different differentiated cells remains absolutely identical, as has been presumed for the last 75 years.

Very powerful tools have been developed, quite recently, that encourage much optimism about our ability to attack these fundamental scientific questions. The most cogent of these is DNA-splicing, or recombinant-DNA technology. This enables us now to isolate specified pieces of DNA from the nucleus of one organism and implant them into a convenient vector, be it a bacteriophage or plasmid. In consequence bits of DNA of specified structure and function can be amplified: that is to say, grown in bulk and treated like chemical reagents.

If we keep in mind that there are probably of the order of 100,000 functional genes in the nucleus of a higher organism and that perhaps 10,000 of these are active in any given cell, we get some conception of the complexity of the task ahead. It is likely that, in order to gain a comprehensive understanding of cell biology, we will need to identify and understand the functioning of at least several thousand gene products, both as individual entities and how they relate to one another. The effect of DNA change, that is genetic variation, on the respective gene products has been one of our most powerful tools for analyzing cell structure and function and will probably play an even more important role in future research.

It is perhaps easier to foresee areas of vital practical application during the next 20 years than it is to anticipate the fundamental discoveries which are always likely to come as a surprise.

Besides its importance as an analytical tool, recombinant-DNA technology is already well on its way to spurring an industrial and medical revolution in the production of specialised biological products like interferon, vaccines, polypeptide hormones, antibodies and other vital proteins. Industrial microbiology of course embraces still larger fields of production of chemicals, antibiotics, amino acids and other essential nutrients. Inexpensive replacements for sucrose, and a host of other large scale chemical industrial processes are being influenced in the most exciting and constructive ways by these new approaches to genetic modiciation of microbial strains. The role that industrial microbiology may play in the development of energy sources is perhaps more problematical. We already have the example of gasohol; and new genetic strains may be expected to play at least a modest role in the conversion of biomass residues that would otherwise present problems of waste disposal. There is little doubt about the economic justification for the application of sophisticated microbial processes when the end-product is expensive, i.e. valued in £ per kilogram or even gram. For cheaper products valued in £ per tonne a more careful appraisal of the economic justification is needed to ensure that the sophisticated techniques are applied to the most fruitful processes for energy and materials conversion on a large scale.

The application of Mendelian genetics to plant breeding has long since demonstrated its immense practical impact in developments like hybrid maize in the United States, and dwarf rice and wheat in the agriculture of developing countries. The fact that for many years farmers in developed countries have faced problems of too much rather than too little grain may have impeded the more urgent application of still more sophisticated scientific methods in plant breeding. Besides the possible extension of DNA-splicing to plant cells we have the already demonstrated approach of fusing somatic cells of plants in culture. Peter Carlson has combined this approach with the selection of desirable properties in cell cultures of the tobacco plant. Although cell fusion still has limited applicability among plant species by current techniques, there can be little doubt that this approach has opened up revolutionary alternatives in the development of plant types for use as crops. As the cost of petrofuel based fertilizers rises; as we become more conscious of other costs of energy in crop production; and particularly as the pressure of the world's growing population on food resources continues its apparently inexorable rise, the need for rapid development of more efficient modes of agriculture becomes obvious. In this sense the development of new crops capable of more efficiently exploiting marginal land resources assumes geo-political significance. This is accentuated by portents of climatic change that may place grave pressures on the agricultural self-sufficiency of the Eurasian land mass.

Somewhat similar remarks can be made about animal production: poultry have been developed that are efficient converters of plant nutrient into meat. However, even more desirable would be the efficient utilization of pasture and other fodders that are inherently unsuitable for human consumption. The task of development of improved breeds of cattle and other large animals with the use of sophisticated genetic approaches is, for obvious reasons, more formidable than poultry development; but large strides have already been made and enormous ones may be anticipated, including the most aggressive exploitation of the existing diversity of germ plasm.

Geneticists face the paradox that the very success of plant breeding tends to drive out wild strains of crop plants. Yet these are a precious and unrenewable resource to provide diverse "germ plasm", needed for disease resistance, productivity under harsh conditions, and other special adaptations. Geneticists are actively involved in field expeditions and in policy and political discussions in efforts to preserve these resources.

More generally, under pressures of economic efficiency, and resource and energy depletion, we need to be able to make more rationally based choices on the balance between the dangers and economic advantages of toxic products of industry and energy generation. Cancer and genetic defects are recognized as the most insidious of these dangers, and genetic methods have already played an important part in the quick screening of chemical substances for their possible environmental hazard.

The reliability of screening methods using bacteria or small animals raises profound questions of comparative biology and genetics. The differences between these species and the human include differences in the metabolism of potentially toxic substances, in immune and repair mechanisms, in many aspects of cell biology. Historically, toxicology has developed as an applied science, which has not in the past used genetic and evolutionary concepts in extrapolating the effects of substances on simpler organisms to estimate their risks to the human. Comparative toxicology presents challenges of the deepest interest to basic science and social policy alike.

If we turn now to the applications of genetics to human affairs, especially to disease: within a few years after the rediscovery of Mendelism in 1900, Archibald Garrod had identified a number of human metabolic diseases which were clearly under genetic control. In fact these studies, which began from a clinical perspective, were the foundations of physiological genetics: that is to say the explanation of gene action in terms of enzymes. For he was able to identify diseases like albinism and alkaptonuria as defects in specific enzymes under genetic control. Since then several thousand genetic syndromes, most of them very rare, have been found and characterised.

In addition, there are a number of unfortunately more common chromosome defects like Downs Syndrome and Klinefelter's. In specific populations, single-gene defects like the thalassemias (southern Europe) and sickle cell disease (among African blacks) are all too prevalent. We have understood for some time that these are a by-product of evolution with heterozygote advantage but have been rather helplessly unable to do very much about it. Around the management of these genetic diseases a sub-profession of genetic counselling has emerged and in the United States this is beginning to be recognized as a professional specialty in its own right. What the genetic counsellor needs to know is how to diagnose specific diseases using cell culture or tissue samples often from amniocentesis, then to provide sensitive counselling to parents who wish to know the prospects of genetic damage and what steps they can take to ensure that they may have healthy children. At issue may be advice about a prospective abortion in the case of a child at risk during pregnancy. This is obviously a grave decision that must be governed by many personal factors, and also by the highest standards of technical realiability about the diagnosis.

New methods for the diagnosis of genetic disease from amniotic fluid cell samples have been developed: the most oustanding recently is Y.W. Kan on the molecular diagnosis of sickle cell disease. For the first time, despite the hullabaloo of a decade ago, it is now possible to advise parents at risk about the prospects that a particular pregnancy will be subject to this highly disabling disease. They may then elect preemptive abortion with a view to ensuring that they can, in further pregnancies, have children with the healthy life-prospect that should be everyone's birthright.

Most single-gene diseases in man are very rare but these have been the only ones which could be carefully studied. There remains the fact that between 20% and 50% of human disease is subject to more subtle variations in propensity or susceptibility. We have had only very primitive methods of studying the genetics of such multifactorial diseases as schizophrenia, heart disease, diabetes or cancer. The Kan methodology now enables the comprehensive mapping of the human chromosome set. From that we can anticipate very rapid leaps to the understanding of the genetic components not only of these diseases but of many other aspects of human longevity and personality. (Much of what's been said until now is hardly more than conjecture because methods like twin comparisons are so unreliable that only the lack of alternative methods can justify their use in research).

Some people are seriously questioning how far we should go in genetic typing for fear of some obvious forms of special abuse. But every advance in genetic insight takes us further and further away from the concept of genetic fatalism and offers more alternatives and options of dealing with the medical problems presented.

Finally, there is little doubt that DNA changes, which is to say genetics at the somatic cell level, are of central importance in the very nature of cancer, ageing and developmental defect. The analytical tools of modern genetics, especially those involving DNA splicing, DNA cloning, the sequencing of DNA, will be our most powerful resources in attacking cancer, heart disease, psychiatric illness: the major scourges of mankind in developed countries at the present time.

The rest, that is most, of the world's population faces a different set of problems that can be summarized as population, famine and plague. Geneticists do not directly intervene in population change, but their skills are indispensable for understanding its consequences for the human condition. The development of new crops is already an important factor in economic development. As to plagues, the most important communicable diseases in the world today are no longer the bacterial and viral infections familiar in developed countries. It is malaria, schistosome and other worm infestations, trypanosomes, Leishmania, and other protozoa, that are the scourges of hundreds of millions of people today. But these very organisms are fascinating objects of basic biological inquiry, presenting a challenge that parallels that of the bacteria a century ago. One of the main frontiers now being reached is the application of the most sophisticated tools of modern molecular biology and genetics to the eukaryotic microbes and worms that are the major public health problems of the developing countries.

Further reading

Judson, Horace, *The Eighth Day of Creation: Makers of the Revolution in Biology*. London: Jonathan Cape, 1979.

Jacob, Francois, *The Logic of Life: A History of Heredity*. London: Allen Lane, 1974.

Whitehouse, H.L.K., *Towards an Understanding of the Mechanism of Heredity*. New York: St. Martin's, 1973.

McCusick, V.A. and Claiborne, R. (eds.) *Medical Genetics*. New York: HP Publ. Co., 1973.

King, R.C. (ed.) *Handbook of Genetics*, Vols. 1-5. London: Plenum, 1974-76.

Vogel, F. and Motulsky, A.G. *Human Genetics*. New York: Springer-Verlag, 1979.

Bodmer, W.F. and Cavalli-Sforza, L.L. *Genetics, Evolution, and Man*. San Francisco: W.H. Freeman, 1976.

The most comprehensive oversight of the field is offered by *Annual Review of Genetics*, published by Annual Revs., Inc., Palo Alto, Calif.

VIROLOGY

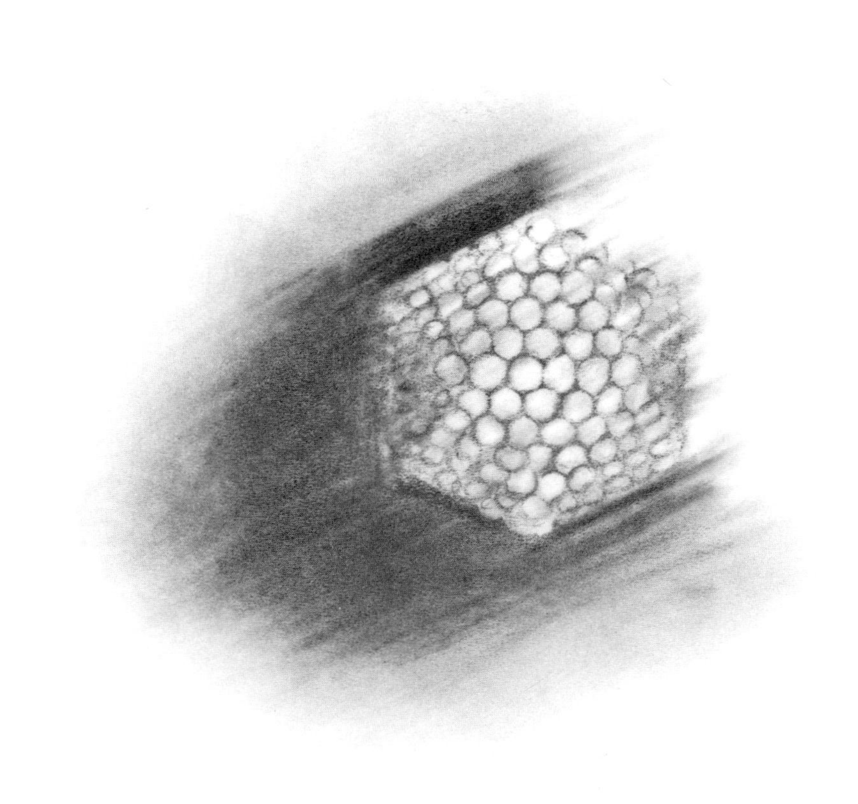

Profile No. 17

VIROLOGY

by

Professor Erling Norrby
(Professor of Virology at the Karolinska Institute, School of Medicine, Stockholm)

Viruses are unique with respect to the way they infect their host cells. They have been the object of many studies because of their well-known capacity to produce disease, and also because they represent the simplest form of self-replicating genetic material. Thus by elucidating the mechanisms by which viruses multiply we can learn a lot about the fundamental characteristics of the molecules of life.

Viruses are parasites in living cells. Two different phases can therefore be distinguished in the life-cycle of a virus. One phase is when the virus is outside the cell, the virus particle. This relatively inert structure is what we normally think about when we use the term virus. The virus particle contains the genetic material of the virus wrapped in a protective shell of protein and in larger viruses also some membrane structure. The genetic material is either deoxyribonucleic acid (DNA) or ribonucleic acid (RNA), but in contrast to other micro-organisms there is no virus which contains both DNA and RNA. The protective shell of protein is built up of subunits in a symmetrical manner which accounts for the fact that the smaller viruses can be crystallized. However, an individual virus particle is very small, less than 0.0003 mm. The finding of virus crystals stirred philosophical discussions on whether viruses represent living or inanimate material. With our present knowledge it is quite apparent that viruses indeed are lively, but this can be recognised only during the second phase of the virus life cycle, i.e. when the virus replicates in infected cells.

There is a very limited amount of genetic material in viruses and some of them can direct the production of only three proteins. Because of this they rely heavily on the normal machinery of cells in which they replicate. Viruses do not replicate by growth and division. Instead the different components of a virus particle are produced separately and then assembled. As many as 100,000 particles can be produced in one cell. The release of these particles frequently is associated with the death of the infected cell.

Viruses can infect vertebrate, insect, plant and bacterial cells. They may cause disease not only in man and animals but also in insects and plants. In fact the first virus disease was identified in tobacco plants with mosaic changes of their leaves. Plant virus particles can frequently be obtained in relatively large quantities, which have facilitated studies of their structure. Recently a new group of virus-like infective agents has been identified in plants. These agents have been called viroids. Viroids are composed exclusively of a small piece of RNA that forms a circular structure. The complete chemical structure of one viroid has been shown. Still nothing is known about the way these infectious agents replicate and how they can cause disease. It is possible that viroids occur also in man and animals. Bacterial viruses, usually called bacteriophages, replicate rapidly in their host cells, which lack a nuclear membrane. Studies of this infectious process have provided much important information on the principles of the interaction between the genetic material of a virus and its host cell.

Virus research can be performed on biological systems at three different levels: (a) detailed analysis of the composition of virus particles and their architecture, (b) studies of virus replication in cells, and (c) identification of changes in the diseased multicellular organism after viral infection.

The genetic material of a virus is one single piece in most viruses. Most often this piece is linear but certain viruses contain a circular form of nucleic acid. A few types of viruses, e.g. influenza virus, have their genetic material divided into segments, with each segment directing the production of one protein. This provides an excellent system for genetic studies. With molecular fingerprinting it is just becoming possible to identify a virus causing a particular disease and epidemic. The complete biochemical composition of both nucleic acid and proteins of some smaller viruses is known today. This knowledge gives fascinating insight into how the information of the genetic material is processed into specific proteins. The regulatory importance of nucleic acid which is not translated into protein is gradually being elucidated. Unexpectedly it was found that one piece of nucleic acid may direct the synthesis of more than one protein by a mechanism for shifting the reading frame. This happens in the following way. The genetic information is coded by the sequence of the four different nucleotide bases in the DNA or RNA. Three adjacent bases code for one of the twenty different amino acids in the polypeptide of a protein. In these terms a reading frame is a triplet of bases so that the sequence of amino acids in the resulting peptide will depend on which base is used to start the reading frame. The general biological significance of this phenomenon is not known.

As was mentioned before, viruses, to a major extent, use the normal cell machinery for their replication. As a consequence much has been learnt about the way in which the genetic material in the normal cell is used for protein synthesis by studies on virus replication. In particular we know that the information contained within the DNA of the cell is used selectively according to the requirements of the cell or the body at any given time. This is known as the regulation of expression and it is here that work on viruses has been particularly valuable. The normal flow of information is DNA → RNA → protein. It is now recognised that some viruses which only contain RNA can transmit their genetic information to the DNA of the host cell. Further it has been found, again by studies of a virus system, that RNA copied from DNA may be modified by a process known as splicing before the RNA is translated into protein. Increasing knowledge about the mechanism of virus replication provides hope that, in the near future, remedies will become available which can prevent virus replication without causing any harm to the infected host cell.

Virus replication does not always lead to the death of cells. In certain cases both the virus and the cell continue to replicate. In other cases the genetic material of the virus associates with the genetic material of cells. As a consequence no virus particles are produced and when the cell divides both daughter cells receive virus genetic material. The virus has become a part of the hereditary material of the cells. This may provide a mechanism for transfer of genetic material within a cell and also between cells. As a consequence normally growing cells may change into tumour cells displaying disorganized growth.

The symptoms of viral diseases result from cell destruction; since a considerable fraction of the cells in a particular organ have to be destroyed before any symptoms occur, most virus infections in man are not associated with symptoms. In response to an acute virus infection, the human body mobilizes defence mechanisms. These often lead to the elimination of the viral infection, but it has also been found that, not infrequently, the virus infection persists. The virus may remain in the body in a quiescent form, for years and even decades; later the virus may become activated and cause disease.

As mentioned previously, only limited advances have been made in the search for antiviral compounds. However, preventative treatment by the use of vaccines has formed the basis of some of the major achievements in the field of medicine. Smallpox has been eradicated from the world, and in many industrialized countries poliomyelitis has been virtually eliminated. New vaccines are continually being introduced.

The study of viruses today occupies a unique position in biomedical and molecular biology research. It spans from the area of sophisticated molecular analyses into the field of highly practical problems, such as the development and application of new vaccines. During recent decades there has been an overwhelming accumulation of knowledge about viruses with a doubling of available information about every fifth year. However, as in many other fields, new knowledge generates an increasing number of unsolved problems. One can therefore predict that virology will maintain its position as one of the most rapidly developing fields of research in biology.

BIOPHYSICS

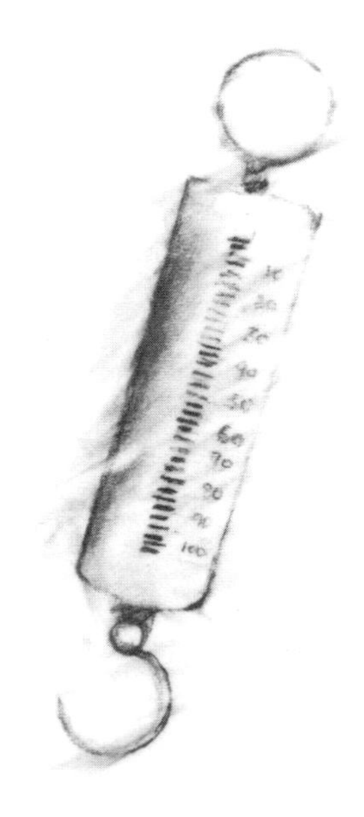

Profile No. 18

BIOPHYSICS

by

Professor David M. Blow

(Professor of Biophysics at Imperial College in the University of London)

I pay little attention to the official labels that are tied on things, and never thought to enquire what the title of my post would be when I came to the Physics Department at Imperial College. I have to admit that I was appalled when a letter informed me that I was Professor of Biophysics. In the same way I had always been sceptical about the title Molecular Biology attached to the laboratory where I worked at Cambridge. That title, which is sure to be discussed elsewhere in this volume, had struck me as a flexible friend, which has been distorted to fit any convenient situation. On reflection, however, I decided that no title would fit the range of activities within my group better than Biophysics.

When I analysed my instinctive reaction to the name, I decided that I had been horrified because chairs of Biophysics at Cambridge, and at University College, are filled by very eminent men who work on the transmission of nerve signals and on the maintenance of electrical potentials across membranes: brilliant work, which has little relation to my own specialities in the fields of large molecule crystallography, and enzyme structure and mechanism. On the other hand, the chairs of Biophysics at King's College and at Leeds, and of Molecular Biophysics at Oxford, are filled by men whose research specialities are much more like my own.

In any case, a definition was clearly needed. To be introduced at a party as a Professor of Biophysics is a solid conversation-stopper; the least one can do is to be prepared for someone who gushes – "Oh, how *exciting*! I've *always* wanted to know about Biophysics. *Do* tell me exactly what it's all about." Also people in the Department, and not least the Physics students who could take it as an optional course, might feel entitled to some kind of an explanation of Biophysics.

The definition which I made up immediately, but which seems to be standing the test of time, is – "The application of physical techniques, and physical methods of analysis, to biological problems."

Physical techniques are more or less transportable to other laboratories. The microscope, the spectrophotometer, the ultracentrifuge, the scintillation counter, the electron microscope, are physical instruments which we expect to see in biological laboratories. They function satisfactorily there, because they are fully developed, self-contained instruments, using physical techniques which are well understood, suitable for application to a wide range of investigations without further development. By contrast, nuclear magnetic resonance, mass spectrometry, X-ray diffraction, stopped-flow fluorescence and dozens of others, are techniques undergoing rapid development, more likely to be pursued effectively if technical innovations can be implemented competently on the spot. In time, I expect these techniques, too, will be exported in the form of standardised, self-contained instruments, for use in the biological laboratory. I cannot predict whether we shall still be working then on rapid computer-controlled image analysis, soft X-ray microscopes, ion-selective microelectrodes and quasi-elastic light scattering for biological applications: more likely, we shall have moved on to other techniques not yet dreamt of.

The phrase "physical methods of analysis" is meant to refer to something deeper – to the contrast in methodology of traditional physics and of traditional biology. Physicists have generally worked from a hypothesis in the form of a *model*: The phenomenon studied behaves as though it were subject to certain rules, rules which generally need to be stated in a mathematical form. Some of these mathematical models, like Newton's definition of force on the basis of mass and acceleration, are so familiar that we accept them without effort. The Lorentz transformation of velocities, which the theory of special relativity employs to make the velocity of light the same to all observers, may seem like a mathematical abstraction which defies common sense.

An acceptable physical model has to be universal – it has to apply everywhere, and it must be consistent with all other physical aspects of the world. Though this is not always achieved, it means that each advance in physics restricts the range of phenomena which remain unexplained, and the range of acceptable physical explanations which may be given for them. In its broader aspects, biophysics examines biological phenomena, determines how they can be fitted into the model system which physics has built up, and draws conclusions about the nature of biological systems.

As in Galvani's experiments on the electrical stimulation of a frog's leg, biological observations have occasionally advanced physics. In the past, physicists have sometimes doubted whether biological systems are subject to physical laws. My favourite example is Lord Kelvin's formulation of the Second Law of Thermodynamics (1854):

> "It is impossible, by means of *inanimate material agency*, to derive mechanical effort from any portion of matter by cooling it below the temperature of the coldest of the surrounding objects."

Doubtless Kelvin was thinking of Maxwell's mythical "demon", a being who guards a frictionless trap-door which separates two gas-filled enclosures, and is equipped with a visual system to see the approaching molecules. He opens the door only when a fast-moving molecule is approaching from one side, or when a slow-moving molecule arrives from the other. In time, one side of the enclosure is filled with slow-moving (cold) molecules, all the energetic (hot) molecules having passed into the other side, and a temperature gradient has been created, apparently without any physical work being done.

The paradox stood for over a century before it was demolished by Brillouin (1956). If the two enclosures are at the same (uniform) temperature, the heat radiation is uniform in all directions. In order to see the approaching molecules, the demon will need a torch, a supply of radiation energy which can be bounced off the approaching molecules into his visual system. By a precise analysis of the minimum energy the demon would dissipate in performing this task, Brillouin was able to show that it matched exactly the work which could be obtained from the temperature difference he had built up.

It is a fundamental tenet of biophysics, as I have described it, that the laws of physics apply to biological systems. The most influential book in the literature of biophysics gives beautiful examples of the application of physical knowledge to place close limits on what is possible in biological systems. Schrödinger's *What is Life?* (1944) was instrumental in persuading many physicists to leave main-stream physics and move into a new, uncharted field: Benzer, Crick and Wilkins were among them. It was partly a matter of timing, for the book appeared at the end of the war when many would have been moving back to academic science, and at the time when the ethical problems created by the Manhattan Project were causing many physicists to re-examine their personal motivation.

What is Life? was published a year after Avery's key experiment on the transforming principle, but Schrödinger apparently did not know anything about DNA. He had, however, gained from Max Delbrück (another renegade physicist) a clear insight into the role of genetics and the nature of the gene. He was able to deduce that a gene, as then defined, could not contain more than a few thousand atoms; he saw with the clearest insight that the gene must be the physical embodiment of a code specification. *What* was encoded, or *how*, was far out of reach (we still only know part of the answer), but the size of the gene, coupled to the thermodynamic limitations of the system, lead deep into the problems of molecular genetics.

When we try to look beyond the equipment and problems of current biophysics, what do we expect to see? I believe firmly that instrumentation is one key to scientific advancement: so that equipment for laser-pulse X-ray diffraction observations, for computerised image analysis in real time, for rapid and reliable separation of different classes of cells, for non-invasive imaging using nuclear magnetic resonance or ultrasound, for direct read-out of the status of nerve cells by optical means, without insertion of an electrode – all these and many more look to me like techniques with which biophysicists should become involved, in order to exploit fully the biological information which they may be able to give.

To look further, to see where the next generation of problems will lead, we need to identify some specific problems. We have learnt so much, on so broad a front, about biomolecular structure and organisation in the last 25 years, that it is easy to forget problems which were obvious throughout that period, but which were laid aside by most of us because there was no obvious line of attack. How does a eukaryotic cell know its identity as a particular cell type? How does the gene specify the spatial organisation of a multicellular organism? What constitutes learning, and where does it take over from evolution? What is memory? These are old questions, and some progress has been made with all of them. I list them now, because they are fundamental questions, basic to all biology, and because physical science lays definite constraints on the types of answer which are possible. It may be too much to hope that they will all be answered in another 25 years, but I certainly hope that some of them will; and I believe that biophysicists will have a part to play in answering them.

BIOCHEMISTRY

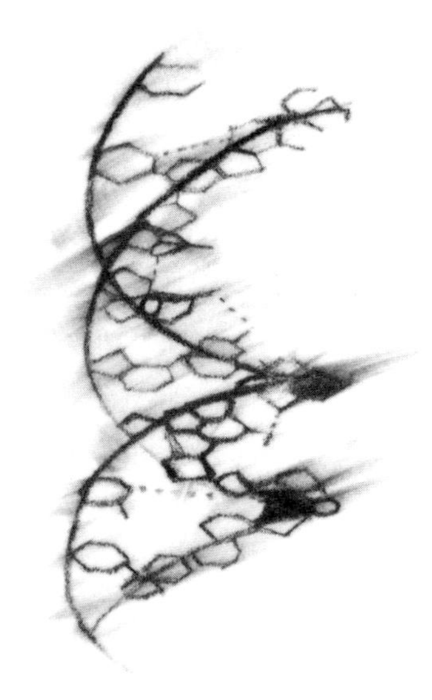

Profile No. 19

BIOCHEMISTRY

by

Professor Emil L. Smith

(Professor of Biochemistry in the University of California, Los Angeles)

A simple definition of biochemistry is that it is the science that deals with the chemistry of living organisms. Biochemistry seeks to explain the properties of living organisms in terms of the behaviour and changes in the chemical substances that are characteristic of these organisms, whether they be simple unicellular ones, or complex plants and animals. If we examine some of these special properties we can approach the study of biochemistry with a better understanding of the fundamental questions it seeks to answer.

What are the characteristic chemical compounds of living things? Which of these are derived from foodstuffs and which are made in each species? How are they made? How is the structure of the cell organized to perform its functions? What is the chemistry of inheritance and how does a cell divide to yield identical daughter cells? How are the thousands of chemical reactions in a single cell or in a complex organism regulated to function in an harmonious way? How does the genetic information in a fertilized egg direct the formation of differentiated, complex organisms? Can behaviour be described in chemical terms? Can disease be understood in molecular terms?

Progress in biochemistry has developed with the sister sciences of chemistry and physics. As these sciences advanced, it became possible to use more and more refined methods to study the structure and behaviour of the characteristic compounds of living matter. Living cells are composed of some of the most complex compounds in existence, the determination of the structure of which presents great technical difficulties. This is not only because many of these substances occur in vanishingly small amounts, but also because hundreds of thousands of reactions are occurring simultaneously. It has been the development of physical instruments and chemical methods of the most sophisticated kind (many devised by biochemists), that now permit some degree of understanding of the nature and behaviour of living organisms.

It is useful to consider first some of the kinds of molecules which are characteristic of living cells. Let us focus on two major groups of substances that have molecular weights in the range from several thousands to many millions. These are known as macromolecules. Although these substances are large, they are all built on a common theme of repeating patterns, utilizing relatively few small molecules as building blocks. First, there are the nucleic acids, more specifically the deoxyribonucleic acids or in abbreviated form, DNA. One of the greatest achievements of modern biochemistry was the recognition that the genetic material is DNA. The familiar facts of genetics are that each species reproduces itself faithfully, and further that in a given species, the offspring resemble the parents.

DNAs are large molecules which, in general, contain only four kinds of smaller molecules, nucleotides, which are linked together. Each nucleotide contains a nitrogen-containing base, the same sugar (deoxyribose) and phosphate. It is through the phosphates that the nucleotides are joined. In each particular DNA, it is the linear order of the four nucleotides, A, G, C, T (in abbreviation) that differentiates it from all other DNA molecules. DNA contains two strands that are held together non-covalently and that are complementary inasmuch as A binds to T and C to G. Thus for the arbitrary sequence for portions of two chains, with the dots indicating the non-covalent binding between the two chains:

```
- - - - - A A A C A T T A G G A T - - -
          . . . . . . . . . . .
- - - - - T T T G T A A T C C T A - - -
```

During cell division, each strand is copied (replication) to form the complementary strand. Therefore, each daughter cell receives a complete and exact set of the two strands of genetic material that was present in the maternal cell.

How then does this genetic material with its precise sequence of nucleotides determine the properties of the cell? One strand of the DNA is copied to form another kind of linear polymer, ribonucleic acid or RNA, which differs from DNA in the kind of sugar, ribose in place of deoxyribose, and which contains the base, U in place of T. Once again the copying (transcription) is done by using the complementarity of bases A→U, G→C, T→A, and C→G. Although there are several kinds of polyribonucleotides, the bulk of the transcribed RNA comprises messengers or messengers RNAs, each of which contains the instructions for the synthesis of the various proteins.

Proteins are polymers made up primarily of amino acids. There are commonly 20 different amino acids in a protein. Each amino acid contains an amino (NH_2) and a carboxyl (COOH) group, H_2N-CH(R)-COOH, in which the side chain or R group differs. In proteins, each amino group is linked to a carboxyl group in an amide (or peptide) bond, —C(=0)-NH—. Proteins may contain from approximately 50 to more than 1000 residues of amino acids in a single chain. Many proteins contain more than one chain which may be identical or different. Since each amino acids can occur more than once in a chain, the possibilities for variations in amino acid sequences are enormous.

A specific messenger RNA (and the corresponding portion of DNA or a gene) contains the code for a single protein chain of amino acids. The coding is accomplished by a system in which a specific triplet of bases in the messenger RNA is read during synthesis for insertion of a single amino acid in the growing protein chain. In addition to specific triplet codes or codons for each of the 20 amino acids, there are codons for starting and stopping the protein chains. Once the chain of amino acids is complete, the protein folds spontaneously to form a complex three-dimensional structure. The protein may be further modified by alterations of the side chains of some of the amino acids or by coupling to various other kinds of molecules — organic structures, metal ions, etc.

It is the proteins that are largely responsible for the functional characteristics of each living cell. The vast majority of proteins are enzymes, complex catalysts which are responsible for specifically directing each of the chemical transformations that occur in a cell. As catalysts, enzymes differ remarkably from other types of catalysts familiar in the chemical laboratory. Each enzyme catalyzes only one specific type of chemical reaction. It does so with amazing efficiency and at the ambient temperature of the cell. Practically all reactions that occur in the living cell are catalyzed by enzymes and do not occur spontaneously. A major advantage to the cell, other than specificity and high yield of product, is that the rate of the enzyme-catalyzed reaction can be regulated by activation or inhibition.

Regulation serves to keep all of the metabolic reactions proceeding in harmonious balance.

Let us think of this in the simplest possible way. For nutrition, a human being requires water, a variety of inorganic ions and about 25 preformed, small organic compounds that are obtained by consumption of plant or animal tissues, as well as substances that are used primarily for energy — mainly carbohydrates and fats. The process is so regulated that when the food contains fats and carbohydrates, the tissues fabricate a balanced supply of all the other small and large molecules needed to make the stuff of the cell itself, including the specific DNA and proteins.

An adult human being will consume over a period of about 40 years approximately 6 tons of solid food and 10,000 gallons (45,500 litres) of water. Yet in the normal individual, there will be little change in body weight or chemical composition of the tissues. Thus, the regulatory processes of metabolism are exquisitely controlled. These control mechanisms operate at the level both of the DNA in regulating the rates at which enzymes and other proteins are made, and at the level of the individual enzymes in supplying just enough chemical energy for the specific synthesis both of the small molecules and the macromolecules that characterize each species. Excess energy is stored as reserve fat and carbohydrate; these are mobilized as needed.

Since the DNA is the genetic material, it is obviously unique not only for a given species but for an individual as well. This is true for the proteins (mostly enzymes) also. Thus, these two types of major polymers must be made within the individual organism and are made from small molecules, nucleotides and amino acids, which are common to all organisms, be they of the simplest or the most complex type. Indeed, even the coding mechanisms for the synthesis of proteins are essentially the same in all organisms.

In addition to the enzymes, various proteins play major roles of many other kinds, particularly in higher organisms. Proteins are involved in immunological defense mechanisms; some serve as hormones; others serve as carriers of oxygen and of small molecules; many are structural elements of cells or organs; some are involved in the contraction of muscles; others participate in a variety of other functions. This is why studies of the chemical reaction mechanisms, regulation and structure of proteins of all types represent a major activity of contemporary biochemistry.

It is indeed one of the truly remarkable findings that one of the major attributes of life — namely, inheritance, can be basically explained by the fact that the information resides in sequences of nucleotides in DNA which can be translated into sequences of amino acids in proteins. The activities of the diverse proteins are then responsible for essentially all of the metabolic behaviour of cells.

Of course, proteins and nucleic acids are not the only important compounds of cells. There are, for example, many smaller compounds called lipids. These include a highly heterogeneous group of compounds whose only common feature is that they are insoluble in water and soluble in organic solvents. Several kinds of complex lipids are major constituents of cellular membranes. Membranes are the devices for separating the contents of a cell from its environment, and for separating the contents of various small organelles (discrete membrane bound components) within the cell from the cytoplasm. Thus membranes serve in maintaining the structural integrity of the cell and its several compartments. In addition to water-insoluble lipids, there are many types of proteins also present in membranes. Membranes are not passive barriers; they also serve in regulating the transport of small and large molecules into and out of cells and their specialized organelles.

The transport of many inorganic ions and small organic ompounds is a dynamic process, inasmuch as the contents of the cell are far from being at equilibrium with the environment. Active "pumping" processes are involved and energy is supplied by the metabolism of small organic compounds.

The metabolism of each cell is characteristic unto itself. Green cells of plants can fabricate all of their organic cellular constituents from carbon dioxide, water and inorganic materials using light energy from the sun. All non-photosynthesizing organisms require some preformed organic compounds in order to obtain energy and to fabricate the macromolecules characteristic of the cell itself. In a simple bacterium each cell carries out all of the functions necessary for survival, growth and reproduction. In higher organisms, there are dozens of differentiated functions: kidney cells excrete, muscle cells contract, nerve cells send messages to different parts of the body, etc. Some cells, like those of the liver, are mainly chemical factories that are responsible for supplying compounds to be used by more specialized cells of other organs and tissues. Nevertheless, all the processes in the highly differentiated and specialized organs are characterized by proteins which carry out the major functions of the tissues.

One of the outstanding problems of present-day biochemistry is to understand how cellular differentiation has taken place. In other words, although each cell has the same complement of DNA and all the specialized cells have arisen from a single cell – the fertilized egg, the adult organism has developed into a complex assemblage of specialized tissues and organs.

Although much of the differentiation process remains unknown, a picture is beginning to emerge. There are enzymes actively functioning in one type of cell which are undetectable in another type of differentiated cell. This means that not all of the information in DNA is being utilized for the synthesis of all enzymes in all cells or at the same time. Thus, the synthesis of proteins must be controlled by activating some part of the DNA to fabricate certain groups of proteins, whereas other parts of the DNA are inhibited and are not being copied. It is thus through activation and inhibition of DNA in the broadest sense that regulation of protein synthesis is achieved. This type of regulation obviously differs from the activation or inhibition of individual enzymes, or the control of passage of ions or molecules into and out of cells.

An example of the latter is our present knowledge of nerve conduction. A nerve cell maintains a voltage difference of 60–70 millivolts across its surface because of an unequal distribution of Na^+ and K^+ ions, a Na^+ pump continuously expelling Na^+ ions. When the potential is initiated, because of a reversal of Na^+-K^+ selectivity, Na^+ ions enter and K^+ ions leave. For a small area of the nerve cell, the resting potential is lowered in the neighbouring area, and the same events occur, and thus the wave of transmission is propagated. The nerve cannot fire during the brief period when the original concentrations of Na^+ and K^+ are being restored. The passage of ions occurs through specific microchannels that are protein in nature.

Each nerve fibre connects with a few or with thousands of other nerve fibres, or with muscle fibres, through minute gaps called synapses. A chemical messenger is released at the end of the nerve fibre and this serves to stimulate the muscle or another nerve by binding to specific protein receptors. Indeed, not only are stimulators released but some nerve endings release inhibitory substances. All of these activators or inhibitors are small organic molecules.

The network of millions of nerve cells and their fibres represents a wiring diagram of the greatest complexity. Yet the basis by which all of the behaviour of the nervous system is achieved is chemical in nature. It should be apparent that the amounts of the neurotransmitters that are released at the ends of nerve fibres are minute indeed. Further, these substances have to be rapidly destroyed in order to prepare the nerve fibres for succeeding messages. Although several neurotransmitters have been identified, others remain to be discovered.

Although the chemical basis for some biological functions is beginning to be understood at least in broad outline, many others still remain rather obscure. A perusal of the contents of this book describing several disciplines of the biological sciences will show how far we still have to go to explain not only the normal processes of all biological systems, but also their relationships to the abnormal, that we call disease. Yet the attributes that characterize human life, growth, reproduction, locomotion, consciousness, etc., all have a chemical basis, and chemistry is the language in which eventually all of these must be understood.

One of the greatest generalizations of biology is the knowledge that all present-day life has a common origin and that the process of continuous evolution over a period of approximately 4000 million years has given rise to the millions of species of organisms that exist today, as well as the even greater number of species that are now extinct. If no other evidence had existed earlier for the evolutionary relationship of all existing species, this is obvious from the facts that the DNAs of all present species contain the same nucleotides and use the same codons for the same 20 amino acids, universally present in proteins. Further, many enzymes and other proteins are present in many different kinds of microorganisms, plants and animals. Examination of the amino acid sequence for a protein obtained from a variety of organisms, shows that much of the sequence has been retained. It follows logically from the concept of evolution, and the evidence also shows, that the more recently two species separated from a common ancestral form, the more similar the sequence, but the greater the time of separation, the greater the divergence in sequence.

Thus, although DNA generally replicates without significant error, some errors are made and changes do occur in the DNA sequence. A few of these errors give rise to defective proteins and these produce metabolic disorders. In human beings, hundreds of such metabolic disorders are now known and in each instance the protein sequence is altered; fortunately, most of these are rare. Some changes in the DNA will be "neutral", that is, there will be no significant effect on protein function. Other changes may actually enhance a desirable property of the protein – produce a more active or more stable enzyme, or alter its specificity with respect to the reactions it catalyzes. When this occurs, the organisms carrying the altered protein may have a selective advantage.

It is commonplace that higher organisms are more complex, thus they have more enzymes, more proteins of other sorts and are more differentiated, and of course, have more DNA. An important feature of evolution has been gene duplication. If one gene continues to make the essential enzyme or other protein, the other gene is free to change. Duplication explains why higher organisms contain more DNA than simpler forms.

Genetic changes in the DNA can take many forms. Alteration of a single nucleotide base results simply in the substitution of a single amino acid in a protein. However, the alterations in DNA may also involve deletions of parts of a gene or whole genes, complex rearrangements, as well as duplication of one or many genes. These changes and others are now being studied by determining the specific sequences of proteins and the responsible genes.

Indeed, with present biochemical techniques and with appropriate enzymes, it is posssible to introduce DNA derived from one species into a bacterial DNA to have the microorganism make the protein of the donor species. Such "genetic engineering" promises much for improvements in food plants, for the production of otherwise unobtainable human proteins for therapeutic use, and for the large scale production of a variety of ordinarily rare enzymes and other proteins for industrial processes.

The application of biochemistry for the improvement of human health and welfare has already made great progress, but the future is even more promising. We are just at the beginning of a whole new era of exciting developments in biochemistry.

MOLECULAR BIOLOGY

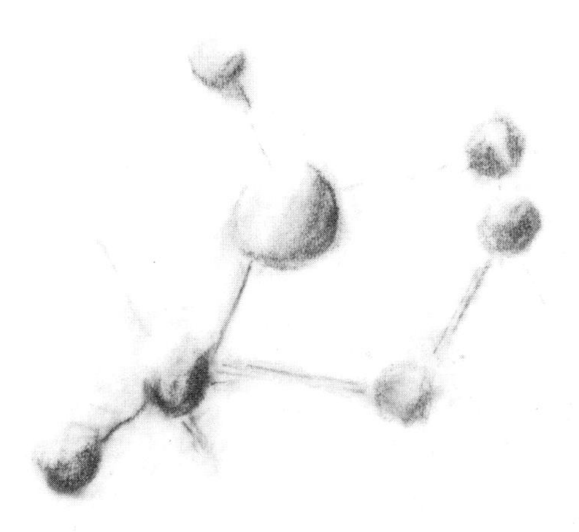

Profile No. 20

MOLECULAR BIOLOGY

by

Professor Peter N. Campbell
(Professor of Biochemistry in the University of London)

If you have read all the other Profiles you will, I hope, have been impressed by the fascinating scope of the biological sciences and the wide range of phenomena which are being described. In any scientific activity the first objective must be to *describe* a phenomenon accurately; one has only to read the Profiles on Ethology or Psychology to see how important and difficult this may be. A further objective of a scientist is to *explain* a phenomenon. This may be achieved at various levels, e.g. animals, organs, cells, etc. You may however have noticed a recurring theme in many of the Profiles where the authors state that a major objective of research in their speciality is to *explain* the phenomena they describe in terms of molecules. This is most explicitly stated under Biochemistry, but it also recurs in the Profiles on Biophysics, Endocrinology, Genetics, Immunology, Microbiology, Pharmacology, Toxicology and Virology. The research scientists who have this common objective in these various fields may call themselves Molecular Biologists. Thus in Endocrinology an important objective is to explain the way in which hormones control the metabolism of their target organs; in Genetics to understand the organization of the genes in the DNA of the cell, in Immunology to explain how antibodies are synthesized in the body as a result of the administration of an antigen. Molecular Biologists aim to explain all these biological phenomena at the molecular level and have devised suitable techniques for their study. Thus the unifying force is derived from a horizontal integration across the various disciplines of biology in contrast to the vertical integration that the various authors of the other Profiles have striven to describe.

You may ask in what way molecular biology differs from biochemistry, which surely has a similar objective. Indeed, since the mid-1950s, when molecular biology first came to the fore, there have been prominent biochemists who have objected to the term "molecular biology", claiming it to be biochemistry in a new guise. While I have much sympathy for that argument I consider it too sweeping. While both the biochemists and the molecular biologists seek to explain biological phenomena in terms of the molecules that are responsible, the coining of the new concept of "molecular biology" attracted a wide range of fine scientists to biology, and they brought with them an entirely new armament of techniques. The concept at the start was mainly promoted by the work of physicists amongst whom Sir John Kendrew deserves special mention. I believe therefore that a major difference between biochemistry and molecular biology has been that the latter has attracted a much more varied group of scientists than would ever have been drawn to biochemistry alone.

Today the actual phrase "molecular biology" is to a large extent confined to organizations and journals. Thus, there is the European Molecular Biology Organization (EMBO) which has had a most powerful influence in engendering enthusiasm for molecular biology in Europe and in attracting finance from the various Governments. There is also now a central laboratory in Heidelberg. The *Journal of Molecular Biology* has also won a fine reputation. There are a few University Departments of Molecular Biology but their number has not increased very much. The learned societies that are associated with molecular biologists, apart from biochemical societies, tend to be the biophysical societies since physicists were attracted to molecular biology.

A key event in the emergence of molecular biology was the publication by Watson and Crick of their views on the structure of DNA in 1955. This gave tremendous impetus to research which had as its objective an understanding of how the biological information in the cellular DNA was utilized to cause the formation of proteins. It should not be forgotten that Sanger at about the same time had determined the structure of insulin. This was the first conclusive demonstration that proteins had indeed a precise molecular structure. The late 1950s, therefore, saw major advances in our understanding of the processes involved in the transcription of the information in DNA to RNA and the translation of RNA into polypeptide chains. The detailed mechanisms of the processes have been the subject of continuing research since then, first with simpler organisms such as bacteria and then with animal cells.

As is described in the Profile on Biophysics, the physicists have been active in many areas of biology, so I will mention only a few of them here. The determination of the three-dimensional structure of proteins by X-ray crystallography has been a major achievement of the molecular biologists. Particular triumphs concern the oxygen-carrying protein haemoglobin and in these discoveries Max Perutz has been specially prominent. Haemoglobin not only has to combine specifically with oxygen under the appropriate conditions in the lungs but also has to release the oxygen in the peripheral tissues. Many genetic variants of haemoglobin have been discovered that affect its detailed structure and change its properties in subtle ways. It is amazing how research in this one field has continued so fruitfully to explain the many properties of this important substance that has evolved to its present sophisticated design.

The work on haemoglobin has shown the way to our understanding of how enzymes, which are themselves proteins, catalyse specific metabolic reactions. Proteins have the very special property of interacting specifically with a wide range of different substances. This is known as "structural complementarity", and today we have a very precise idea of how at least some enzymes catalyse reactions.

Another area that has been of particular interest to the physicists and the physiologists is that of muscle contraction. Here, we have the challenge of determining how chemical energy can be converted into physical movement. The expertise of the electrical engineer, the physicist, the electron microscopist, the physiologist and the biochemist have been brought to bear on this problem, and have brought major advances in our understanding.

Encouraged by their past successes, molecular biologists have more recently turned to even more challenging subjects. A most popular field is neurobiology. It has proved possible at the molecular level to isolate specific protein receptors for substances which excite nerves to respond to stimuli. Such receptors are present in skeletal muscle and electric organs. Others have entered the even more challenging field of behavioural studies.

You may have noticed among the Profiles reference to "recombinant DNA technology" or "genetic engineering". As has been explained, it is now possible to insert a gene from, say, an animal cell into a bacterium and so produce a new strain of bacterium that has the potential to synthesize a much wanted human protein. These developments have generated a new enthusiasm for molecular biology among industrialists and politicians and the new word "biotechnology" has been coined.

At the fundamental level the molecular biologist selects the phenomenon on which he wishes to work and then uses the most appropriate living tissue, whether it be a bacterium or the electric organ of an eel. In a more applied field he may be working with experts in fermentation technology. In any event the research groups involved in molecular biology are likely to be multidisciplinary and the encouragement of the formation of such groups by the molecular biologists has in itself been a major contribution to biology.

I should also say something about the approach of the molecular biologist to the interpretation of experimental results. Because of the immensity of the challenge of many of the phenomena of biology it has often not been possible to make a direct approach to some problems. The older chemist was never really satisfied until he had obtained a substance in a crystalline state and had shown that it had a sharp melting point. The molecular biologist will approach a problem from many different directions, none of which can alone give a definitive answer but together they combine to give a reasonable probability of correctness. Thus, in trying to find the precise site of protein synthesis in a cell, we may disrupt the cell and use a centrifuge to separate the particles according to their density, size and shape. One can then discover whether such particles are able to synthesize protein when incubated in a test-tube. How does one know that the particles isolated are not artefacts produced during the disruption of the cell? Why not examine thin sections of the cell by electron microscopy to see whether one can detect such particles in the intact cell? Even then the critics could claim that the preparation of the sections for electron microscopy had created artefacts so another approach must be devised using the intact animal. It will be seen therefore that a whole variety of methods and technology has been brought to bear on the problem. This is one of the fascinations of the subject but it calls for people with a very broad range of talents. First, they must be interested in the phenomena of biology and then they must have the ability to take a quantitative approach to a problem. This calls for a sound knowledge of physics and chemistry.

Probably, the easiest entry to molecular biology is a university course in biochemistry combined with chemistry and physics in the early stages. Later, it would be good to study genetics, microbiology and virology. But in the end the strength of the subject has been that it has attracted good scientists from many fields, so it is futile to try to lay down a precise training programme.

Indeed, it is appropriate that this volume should end with a reminder that the future of biological research depends on its attracting the finest scientists wherever they may be in the world. It is a help if one has had a sound preliminary education at the hands of the best teachers, but those in positions of responsibility must surely do all they can to encourage people who can bring to the subject original concepts and methods of experimentation irrespective of the nature of their formal training. While in a sense molecular biology has failed to establish itself as a discipline like the other subjects of these Profiles, so that today one does not flaunt the fact that one is a molecular biologist rather than, say, a biochemist, this does not belittle its contribution to biology. Indeed, it was because of its contribution to the unification of biology that I thought it appropriate that the Profile on Molecular Biology should come at the end of the volume.